D0318930

CONTROL
ENGINEERING

W Bolton

Longman
Scientific &
Technical

Longman Scientific & Technical
Longman Group UK Limited
Longman House, Burnt Mill, Harlow,
Essex CM20 2JE, England
and Associated Companies throughout the world.

Copublished in the United States with
John Wiley & Sons, Inc., 605 Third Avenue, New York, NY 10158

© Longman Group UK Limited 1992

First published 1992

ISBN 0 582 097290

British Library Cataloguing in Publication Data
A catalogue record for this book is available from the British Library

Library of Congress Cataloging-in-Publication Data

Set by 8 in Linotron Times 10/12 pt

Printed in Malaysia !

Contents

Preface

This book aims to help the reader:

1 appreciate the basic concepts of open-loop and closed-loop control systems;
2 develop models for mechanical, electrical, fluid and thermal systems;
3 use the modelling approach to the analysis of first- and second-order systems;
4 determine the responses of systems to inputs by deriving and solving first- and second-order differential equations;
5 determine the responses of systems to step, impulse and ramp inputs using the Laplace transformation;
6 construct system models using block diagrams;
7 determine steady-state errors for systems;
8 use pole-zero methods to analyse the behaviour of systems and stability;
9 appreciate the functions that can be used for controllers;
10 determine the frequency response of systems, including the use of Bode and Nyquist diagrams;
11 appreciate the functions of elements in digital systems, including an introduction to z-transforms.

Each chapter includes worked examples and problems, answers being given to all.

The book is seen as being particularly suitable for HNC/NND technicians, more than covering block M of the BTEC bank of objectives Electrical and Electronic Principles N U86/329 and blocks J and P of the BTEC bank of objectives Electrical and Electronic Principles H 13683B. It is also seen as being of relevance to students on other courses where there is a serious introduction to control system analysis for engineers. The level of mathematics assumed is a basic grounding in algebra and calculus.

W. Bolton

1 Control systems

Introduction

What is a 'system' and, in particular, what is a 'control system'? A system can be thought of as just a black box which has an input and an output. It is a black box because we are not concerned with what goes on inside the box but only the relationship between the output and the input. It is a control system if the output is controlled so that it is made to have some particular value or to change in some particular way. Thus a central-heating control system is designed to control the temperature to some specific value, while a machine tool may be controlled to follow a particular set path. This book is about control systems and this first chapter can be considered to be a superficial overview of the basic forms that can be taken by control systems with more depth, i.e. more realistic models, being given in the other chapters of the book.

Systems

The term *system* is used to describe a collection of interacting components around which an imaginary boundary can be drawn so that we can be concerned just with the input or inputs to the system and its output or outputs without the need to explore the detailed interactions between the constituent components. The important aspect of a system is the relationship between the inputs and outputs. A system can be an entire power station or perhaps just an electric motor. However complex the collection of components and their interactions within the system we can effectively consider all the components to be located within a box and all we have to consider are the inputs and outputs to the box. Figure 1.1 shows how we can thus represent a system by a box with the inputs and outputs to the system being indicated by arrowed lines, the directions of the arrows indicating whether they refer to an input or an output. Thus Fig. 1.1(a) shows the power station system with its input of fuel and its output of electricity, Fig. 1.1(b) being the electric motor with its input of

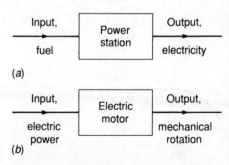

Fig. 1.1 Systems: (a) a power station, (b) an electric motor

electrical power and its output of mechanical rotation.

The advantage of studying systems in this way is that though there is a wide variety of systems possible, the relationships between the output and input of many systems tend to be similar. Thus, for example, the response of an electrical system consisting of a capacitor in series with a resistor to a sudden application of a voltage follows the same type of relationship to the response of a container of liquid to a sudden application of a heat input (Fig. 1.2). Thus, by studying a model system with this type of relationship between input and output we are able to determine how many different forms of system with the same output-input relationship will respond.

In some situations it is convenient to break a system down into a series of linked subsystems. Thus, for example, we might have a temperature-measurement system which consists

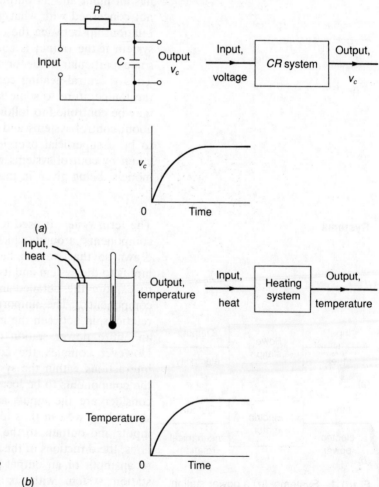

Fig. 1.2 Similar systems: (a) the CR system, (b) the heating system

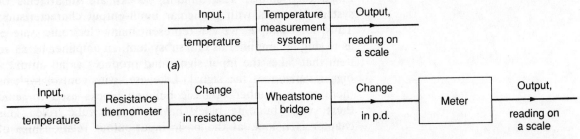

Fig. 1.3 A temperature-measurement system (a) and its subsystems (b)

Models

Fig. 1.4 A central heating system

of a resistance thermometer connected to a Wheatstone bridge with the output being displayed on a meter. We could represent the entire system as having an input of temperature and an output of a reading on a scale, as in Fig. 1.3(a), or as a resistance thermometer subsystem connected to a bridge subsystem connected to a meter system, as in Fig. 1.3(b).

A *control system* is one where the output of the system is controlled to be at some specific value or to change in some prescribed way as determined by the input to the system. Thus a temperature-control system, e.g. the central heating system in a house (Fig. 1.4), could have as its input the thermostat or control panel being set to the temperature required and its output the actual temperature produced. This temperature is adjusted by the control system so that it becomes the value set by the input to the system.

A model ship is a scaled-down version of the full-sized ship, a model aeroplane a scaled-down version of the full-sized plane. The models have the same types of relationships between the lengths of different parts as those in the full-size items. A map is a model of the countryside, the distances and locations of town names on the map having the same types of relationship as occur with the towns in the countryside. A *model* is just a means of transferring some relationship from its actual setting to another setting. In carrying out the transfer we only transfer those relationships which are of interest. Thus the map is only used to transfer relationships involving distances and hence locations, it does not transfer the smells or noises of the countryside.

Drawing a box with arrows for the input and output is just model making for a system. The relationships that are transferred from the actual system to the drawing are the input-output relationships. No one is suggesting by the drawing that the system actually looks like a box with arrows.

A child's construction set of building blocks can be used by the child to make models of houses, cars, cranes, etc. A number of models can thus be constructed from a basic kit. With systems we can similarly use a basic kit to construct a range of systems. The building blocks are subsystems or system elements with particular input-output characteristics. Thus, for example, we can represent many electronic systems as having an amplifier as one subsystem, an amplifier being an item that takes the input signal and produces as an output a bigger version of the signal. Likewise, with control systems there are a number of basic building blocks used to build them, each building block having a particular role. This chapter is a look at the basic input-output relationships of control systems and the roles of the constituent building blocks.

Open- and closed-loop systems

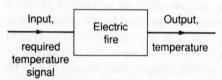

Fig. 1.5 An example of an open-loop control system

There are two basic forms of control system, one being called *open-loop* and the other *closed-loop*. With an open-loop system an input to a system is chosen on the basis of experience of such a system to give the value of the output required. This output, however, is not modified to take account of changing conditions. Thus, for example, an electric fire (Fig. 1.5) might have a selection switch which allows a 1 kW or a 2 kW heating element to be selected. The input to the system is thus a signal determined by whether the switch is set to the 1 kW or the 2 kW setting. The temperature produced in a room heated by the heater is only determined by the fact the 1 kW element was switched on and not the 2 kW element. If there are changes in the conditions, perhaps someone opening a window, the temperature will change because there is no way the heat output can be adjusted to compensate. This is an example of open-loop control in that there is no information fed back to the element to adjust it and maintain a constant temperature. Control systems operated by preset timing mechanisms are open-loop systems.

With a closed-loop system a signal is fed back to the input from the output and used to modify the input so that the output is maintained constant regardless of any changes in conditions (Fig. 1.6). The heating system with the electric fire could be made a closed-loop system if someone with a thermometer monitors the temperature in the room and switches the 1 kW and 2 kW elements on or off to maintain the temperature of the room constant. In this situation there is feedback of a signal to the input related to the temperature, the input to the system then being adjusted according to whether its output is the required temperature. The input to the heater thus depends on the deviation of the actual temperature from the required temperature.

To illustrate further the differences between open- and closed-loop systems, consider a motor. With an open-loop system the speed of rotation of the shaft might be determined

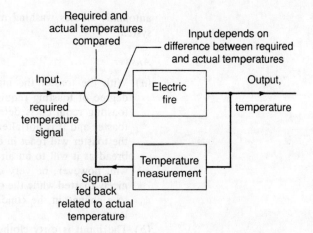

Required and
actual temperatures
compared

Input depends on
difference between required
and actual temperatures

Input,
required
temperature
signal

Electric
fire

Output,
temperature

Temperature
measurement

Signal
fed back
related to actual
temperature

Fig. 1.6 An example of a closed-loop
control system

solely by the initial setting of a knob which affects the voltage applied to the motor. Any changes in the supply voltage, characteristics of the motor as a result of temperature changes, or shaft load will change the shaft speed and not be compensated for. There is no feedback loop. Whereas with a closed-loop system the initial setting of the control knob will be for a particular shaft speed and this will be maintained by feedback, regardless of any changes in supply voltage, motor characteristics or load. In an open-loop control system the output from the system has no effect on the input signal. In a closed-loop control system the output does have an effect on the input signal, modifying it to maintain an output signal at the required value.

Open-loop systems have the advantage of being relatively simple and consequently low cost with generally good reliability. However, they are often inaccurate since there is no correction for error. Closed-loop systems have the advantage of being able to match the actual to the required values. However, problems can arise if there are delays in the system. Such delays cause the corrective action to be taken too late and can, as a consequence, lead to oscillations of the input and instability (see later this chapter). Closed-loop systems are more complex than open-loop systems and so more costly with a greater chance of breakdown as a consequence of the greater number of components. See later this chapter for discussion of the advantages of closed-loop systems with regard to minimizing the effects of changes in the input-output relationships of system elements as a result of environmental changes and the effects of disturbances on the system.

Example 1

Identify the overall output and input and suggest the type of control system that might be used with (*a*) an automatic bread toaster, (*b*) an automatic clothes washing machine, (c) a domestic central heating system?

Answer

(a) With the toaster the input is bread and instructions as to the degree of toasting required, the output is toast. The degree of toasting required is determined by adjusting the setting of the toaster and it is not altered by the condition of the bread. Thus the toaster will react in exactly the same way to a fresh piece of bread as it will to an already toasted piece of bread, the output will, however, be very different with one piece of bread being nicely toasted while the other is black. The toaster does not react to a change in the condition of the bread. The system is open-loop.

(b) The input is dirty clothes and settings of the control dials and switches on the machine for the type of material and form of wash required, the output is clean clothes. The automatic washing machine is open-loop since the machine will carry out exactly the same cycle of washing procedures regardless of whether you put dirty or clean clothes into it.

(c) The input is the required temperature and the output the actual temperature. The domestic central heating system is closed-loop since a thermostat is used to ensure that the input is adjusted to take account of changes in conditions and so maintain a constant temperature.

Basic elements of an open-loop system

An open-loop system can be considered to consist of a number of basic subsystems arranged as shown in Fig. 1.7. These items may not be distinct, separate, items of hardware but all the functions of the subsystems will be present. The overall input to the system is a signal which, on the basis of past experience, is likely to lead to the required output. The subsystems are:

1 *Control element* This element determines what action is to be taken in view of the input to the control system.

2 *Correction element* This element responds to the input from the control element and initiates action to change the variable being controlled to the required value.

3 *Process* The process or plant is the system of which a variable is being controlled.

Fig. 1.7 The subsystems in an open-loop control system

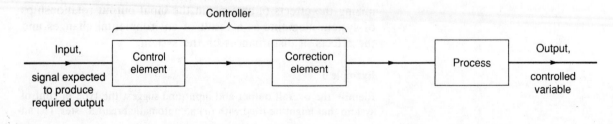

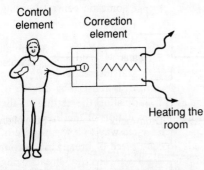

Fig. 1.8 A room-temperature open-loop control system

The first two subsystems above are often combined into an element called a *controller*.

An example of an open-loop system is an electric fire which is used to heat a room (Fig. 1.8). With such a system we have:

Controlled variable – room temperature

Control element – a person making decisions based on experience of the temperatures produced by switching on the fire

Correction element – the switch and the fire

Process – the room

Many open-loop control systems use a control element which sends a signal to initiate action after some time interval or a sequence of signals to initiate a sequence of actions at different times. The controller in such systems is thus essentially a clock-operated switching device. An example of such a control system is the basic operating cycle of the domestic washing machine (Fig. 1.9). The sequence might be:

1 Set controls for the type of clothing being washed.
2 Switch on and start clock.
3 Fill with cold water, the valve allowing water to enter being opened for a specific time.
4 Heat water, the heater being switched on for a specific time.
5 Wash, the drum of the washing machine being rotated for a specific time.
6 Empty water, the valve being opened at a specific time.
7 Fill with cold water, the valve allowing water to enter being opened for a specific time.
8 Rinse, the drum of the washing machine being rotated for a specific time.
9 Empty water, the valve being opened at a specific time.
10 Spin, the drum of the washing machine being rotated for a specific time.
11 Stop, after a certain time has elapsed.

In addition to the above open-loop control system the washing machine is likely to have a number of other control systems for safety, e.g. water level and temperature systems which can switch the system off if the water or temperature rises too high.

Example 2

Identify the subsystems in an open-loop controlled speed motor.

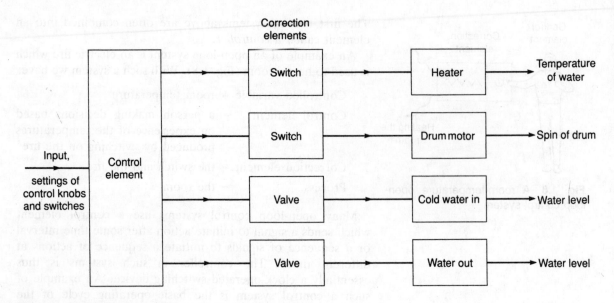

Fig. 1.9 The domestic washing machine

Answer

Controlled variable – motor speed

Control element – a person making decisions based on experience of the speeds produced by switching on the motor

Correction element – the switch

Process – the motor

Basic elements of a closed-loop system

A closed-loop system can be considered to consist of a number of basic subsystems arranged as shown in Fig. 1.10. These items may not be distinct, separate, pieces of hardware but all the functions of the subsystems will be present. The overall input to the control system is the required value of the variable and the outcome is the actual value of the variable.

1 *Comparison element* This compares the required or reference value of the variable being controlled with the measured value of what is being achieved and produces an error signal which indicates how far the value of what is being achieved is from the required value.

Error signal = reference value signal

– measured value signal

2 *Control element* This element decides what action to take when it receives an error signal. The term *controller* is often used for an element incorporating the control element and the correction unit.

3 *Correction element* This element is used to produce a

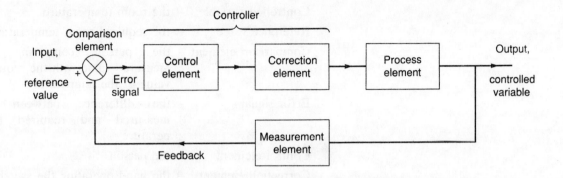

Fig. 1.10 The subsystems in a closed-loop control system

change in the process to remove the error and is often called an *actuator*.

4 *Process element* The process, or plant, is the system of which a variable is being controlled.

5 *Measurement element* This produces a signal related to the variable condition being controlled and provides the signal fed back to the comparison element to determine if there is an error.

A necessary feature of a closed-loop control system is a *feedback loop*. This is the means by which a signal related to the actual condition being achieved is fed back to be compared with the reference signal. The feedback is said to be *negative feedback* when the signal which is fed back is subtracted from the reference value, i.e.

Error signal = reference value − feedback signal

Negative feedback is necessary for control to be achieved. *Positive feedback* occurs when the signal fed back is added to the reference value, i.e.

Error signal = reference value + feedback

In Fig. 1.10 the feedback signal is combined with the reference value at the comparison element. The comparison element is indicated by a circle with a cross, this being the general symbol for a summation element. With the comparison element there is negative feedback and so the reference value is marked as a positive signal and the feedback signal as negative so that the output from the summation element is the difference between the signals. If there was positive feedback at a summation element then both signals would be marked as being positive.

As an illustration of the above discussion of control system elements, consider the control system discussed earlier with Fig. 1.6 where the temperature of a room was controlled by a person switching on or off an electric fire according to whether the room temperature as indicated by a thermometer was at the required value (Fig. 1.11). The various elements of this control system are:

Controlled variable	– the room temperature
Reference value	– the required room temperature
Comparison element	– the person comparing the measured value with the required value of the temperature
Error signal	– the difference between the measured and required temperature
Control element	– the person
Correction element	– the hand operating the switch on the fire and so the heating element
Process	– the room
Measuring device	– a thermometer
Feedback	– negative

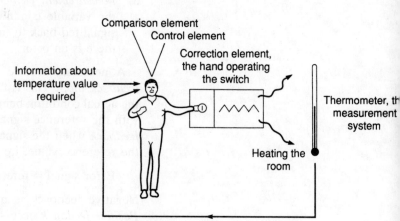

Fig. 1.11 A room-temperature closed-loop control system

Example 3

The domestic toaster is an open-loop control system (see Example 1), suggest a means by which it could be made a closed-loop control system.

Answer

For the toaster to be a closed-loop system there must be a feedback signal to indicate the degree of browning of the toast. Possibilities could be a person watching the toast or perhaps a photocell which responds to the degree of browning. The output from either of these 'measurement systems' would be a signal which is subtracted from the reference signal used to specify the degree of browning required. This error signal could then be used to actuate a relay to switch the toaster heater on or off or a potentiometer to vary the voltage applied to the heater. Figure 1.12 shows the form of such a control system.

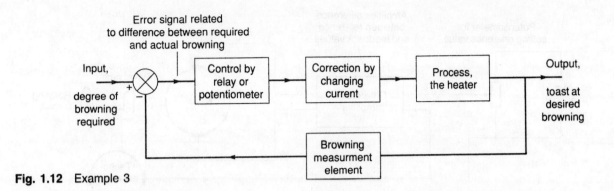

Fig. 1.12 Example 3

Examples of closed-loop control systems

Figure 1.13 shows an example of a simple control system used to maintain a constant water level in a tank. The reference value is the initial setting of the lever arm arrangement so that it just cuts off the water supply at the required level. When water is drawn from the tank the float moves downwards with the water level. This causes the lever arrangement to rotate and so allow water to enter the tank. This flow continues until the ball has risen to such a height that it has moved the lever arrangement to cut off the water supply. It is a closed-loop control system with the elements being:

Controlled variable – water level in tank

Reference value – initial setting of lever position

Comparison element – the lever

Error signal – the difference between the actual and initial setting of the lever positions

Control element – the pivoted lever

Correction element – the flap opening or closing the water supply

Process – water in the tank

Measuring device – the floating ball and lever

Feedback – negative

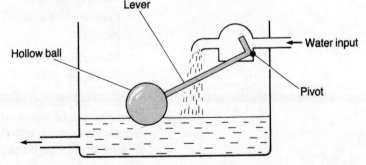

Fig. 1.13 The automatic control of water level in a tank

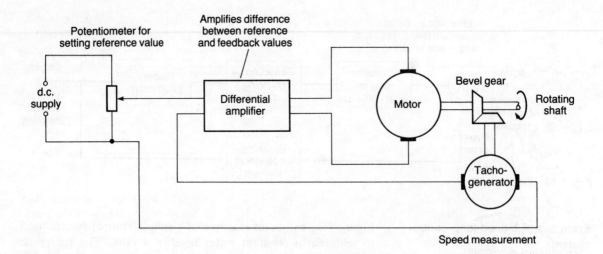

Fig. 1.14 Automatic shaft speed control

Figure 1.14 shows a simple automatic control system for the speed of rotation of a shaft. The potentiometer is used to set the reference value, i.e. what voltage is supplied to the differential amplifier as the reference value for the required speed of rotation. The differential amplifier is used to both compare and amplify the difference between the reference and feedback values, i.e. it amplifies the error signal. The amplified error signal is then fed to a motor which in turn adjusts the speed of the rotating shaft. The speed of the rotating shaft is measured using a tacho-generator, connected to the rotating shaft by means of a pair of bevel gears. The signal from the tacho-generator is then fed back to the differential amplifier. Thus the system has:

Controlled variable	– speed of rotation of shaft
Reference value	– the voltage setting for the required speed
Comparison element	– the differential amplifier
Error signal	– the difference between the reference value voltage and the feedback voltage
Control element	– amplifier
Correction element	– the motor
Process	– the rotating shaft
Measuring device	– the tacho-generator
Feedback	– negative

There are many simple control systems in everyday life. Thus the act of trying to pick up a cup of coffee from a table requires a control system with feedback. The hand doing the picking up has to be moved to the right position, negotiating any obstacles in the way, and then positioned correctly so that the fingers grasp the cup handle in just the right way so that it can be lifted. To control the hand so that it carries out these tasks the feedback used is sight and touch (Fig. 1.15). Thus for the control system for moving the hand to the location of the cup we have:

Controlled variable	– location of the hand relative to the cup
Reference value	– the cup location
Comparison element	– the person
Error signal	– the difference between the actual and required locations of the hand
Control element	– the person
Correction unit	– the arm and wrist
Process	– the hand dynamics
Measuring device	– visual observation
Feedback	– negative

With many control systems there is just a single variable that has to be controlled, e.g. the level of water in a tank or the speed of rotation of a shaft or the location of a hand. There are however some control systems where more than one variable has to be controlled. The act of picking up the coffee cup is such an example since not only has the location of the hand to be controlled but also the pressure exerted by the fingers in grasping the cup handle. There are thus two feedback loops, one concerned with location and using visual means of measurement and one concerned with pressure and using touch as a means of measurement (Fig. 1.16). This can

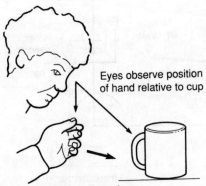

Eyes observe position of hand relative to cup

Arm and wrist movements used to correct hand location

Fig. 1.15 Moving to pick up a cup of coffee

Fig. 1.16 Picking up a cup of coffee

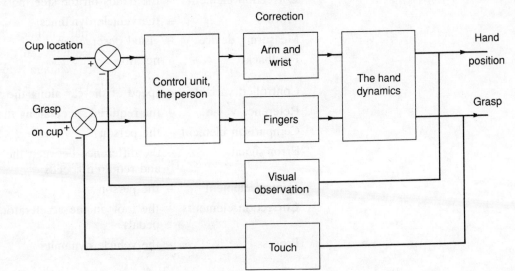

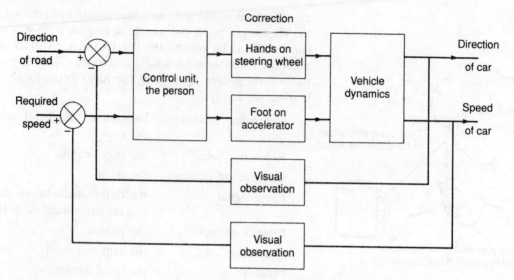

Fig. 1.17 Driving a car

also be considered to be the basic form of a robot arm with a gripper designed to pick up objects on a production line.

Driving a car is another example of a control system where there are two variables that have to be controlled, namely the direction of the car and its speed (Fig. 1.17).

Controlled variable	– direction of the car along the road
Reference value	– the required direction along the road
Comparison element	– the person
Error signal	– the difference between the actual and required directions
Control element	– the person
Correction element	– the hands on the steering wheel
Process	– the vehicle dynamics
Measuring device	– visual observation
Feedback	– negative

Controlled variable	– speed of the car along the road
Reference value	– the required speed along the road
Comparison element	– the person
Error signal	– the difference between the actual and required speeds
Control element	– the person
Correction element	– the foot on the accelerator/brake pedals
Process	– the vehicle dynamics

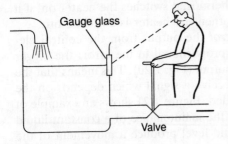

Fig. 1.18 Example 4

Measuring device   – visual observation

Feedback   – negative

Example 4

A workman maintains the level of the liquid in a container at a constant level. He does this by observing the level of the liquid through a gauge glass in the side of the tank and adjusting the amount of liquid leaving it by opening or shutting a control valve (Fig. 1.18). For such a control system what is (*a*) the controlled variable, (*b*) the reference value, (*c*) the comparison element, (*d*) the error signal, (*e*) the control element, (*f*) the correction element, (*g*) the process, (*h*) the measuring device?

Answer

Controlled variable – level of liquid in the tank

Reference value – the required level, probably indicated by some mark on the glass

Comparison element – the person

Error signal – the difference between the actual and required levels

Control element – the person

Correction element – the valve

Process – the water in the container

Measuring device – visual observation of the gauge glass.

The control element has as its input the error signal and as its output a signal which becomes the input to the correction unit so that action can be initiated to eliminate the error. There are a number of ways by which a control element can react to an error signal.

With open-loop control the types of control are likely to be on-off switching or timed switching sequences or actions. An example of the on-off switching control is a person switching on an electric fire to obtain the required temperature in a room. An example of a timed switching sequence is the operation of a domestic washing machine (see earlier this chapter for details).

With closed-loop control the types of control are often two-step mode control, proportional control or proportional combined with some other refinements. With *two-step mode control*, the error signal input to the control element results in either an on or an off output which is used to switch the correction element on or off (Fig. 1.19). Thus in the case of a domestic central heating system controlled by a thermostat, the thermostat gives an output which switches the heat on or off depending on the error. If the room temperature falls below

Control strategies

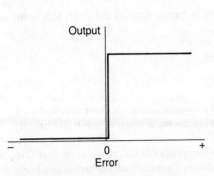

Fig. 1.19 Two-step mode control

a certain value then the thermostat switches the heater on, if it rises above the set value then the heater is switched off.

With *proportional control* the output from the control element is a signal which is proportional to the error: the bigger the error the greater the output (Fig. 1.20). This means that the correction element will receive a signal which depends on the size of the correction needed. Figure 1.21 shows an example of such a control system for the maintenance of a constant liquid level. Changes in the liquid level produce a movement of the float and hence the pivoted arm to which it is attached. This in turn changes the opening of a valve and so affects the rate at which liquid leaves the tank. The bigger the error in the liquid level the greater the change in the valve opening.

Because proportional control alone can present a number of problems, it is often combined with other forms of control. These are *derivative control*, where there is an output proportional to the rate of change of the error signal, and *integral control*, where there is an output at a time t proportional to the integral of the error signal between $t = 0$ and t. A simple example of proportional control plus derivative control is an automatic vehicle where the controller takes action based on an assessment not only of the position of the vehicle but its speed, i.e. rate of change of distance. With just the proportional control the controller is just giving a response in proportion to the size of the error from the required position. It does not take into account how fast the error is changing. Derivative control does. Thus if the vehicle started to move rapidly away from the required path, with derivative control there would be a bigger corrective action than if the vehicle only slowly moved away from the required path. The combination of proportional plus derivative control is thus more quickly able to take account of deviations from the required path and correct them. The above forms of control strategies are discussed in more detail in Chapter 8.

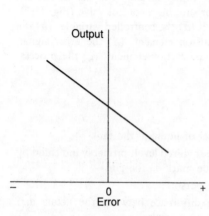

Fig. 1.20 Proportional control

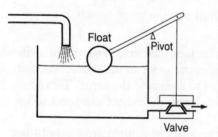

Fig. 1.21 Proportional level control system

Example 5

Which type of control strategy is being exercised in the following control systems?
(*a*) A refrigerator.
(*b*) Steering a car along a road.

Answer

(*a*) With the refrigerator the temperature is controlled by a thermostat and is thus likely to be a simple two-step, on-off, control system.
(*b*) This is basically likely to be proportional control in that the controller output, i.e. the driver's rotation of the steering wheel, is likely to be proportional to the error.

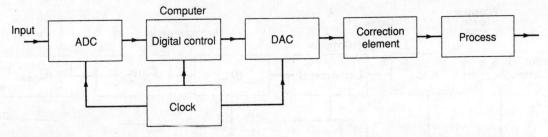

Fig. 1.22 Continuous time open-loop digital control system

Digital control

Digital control by means of microprocessors is being used increasingly in control systems. Figure 1.22 shows the basic form that can be taken by an open-loop digital control system when the input is a so-called continuous time signal, i.e. a signal that can be varied continuously with time and is not a digital input signal. The input signal first passes through an analogue-to-digital converter (ADC). This converts it into a digital signal, a coded number. The digital control element incorporates a clock which sends out pulses at regular time intervals. Every time the ADC received a pulse it sends the coded number to the digital control element. The digital control element implements a control strategy determined by a stored program. Because such a program can easily be changed the control strategy can also easily be changed and this flexibility is a great advantage over analogue control systems where the control strategy is determined by pieces of hardware. The signal output from the control element is also in digital form, a coded number. It is converted into an analogue signal by a digital-to-analogue converter (DAC) so that it can actuate the correction element and so produce the required change in the process variable.

The sequence of steps that take place in the four blocks that constitute the control element are:

1   Wait for a clock pulse.
2   Carry out an analogue-to-digital conversion of the input signal at that time.
3   Compute the control signal according to the control strategy in the program.
4   Carry out a digital to analogue conversion of the control signal.
5   Update the state of the correction unit.
6   Wait for a clock pulse and then repeat the above cycle.

Since the input is only sampled at certain times, such a system is often referred to as a *sampled data system*. Sampling is a fundamental aspect of digitally controlled systems.

Figure 1.23 shows the form that can be taken by a

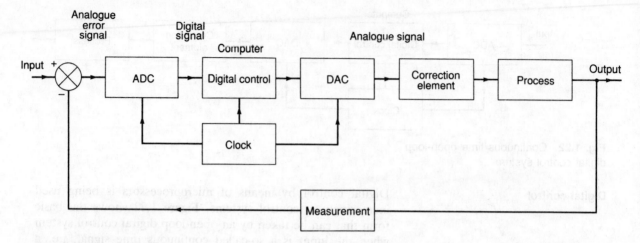

Fig. 1.23 Continuous time closed-loop digital control system

continuous time closed-loop digital control system. The analogue error signal from the comparison element is converted into a digital error signal by the analogue-to-digital converter (ADC), acted on by the digital control element according to the control strategy program and then converted into an analogue signal for the correction element by the digital-to-analogue converter (DAC). The sequence of steps that take place in the control element are as with the open-loop system described above.

A single digital computer can be used to control several different variables. For example, the domestic washing machine described in Fig. 1.9 could use a microprocessor as the control element. Digital control systems are used in many applications, e.g. machine tools, aircraft control, and robots. For a more detailed discussion of digital control systems see Chapter 9.

Mathematical models for systems

In order to be able to understand the behaviour of systems it is necessary to obtain mathematical models of them. A model ship is a scaled-down replica of the full-sized ship. In the model the relative sizes of the various parts are in the same proportions as in the full-size ship, i.e. there is a constant scaling down of the sizes. A photograph can be considered to be a model of the scene that was photographed. A *mathematical model* of a system is a 'replica' of the relationships between the input and output or inputs and outputs. The actual relationships that exist between the input and output of a system have been replaced by mathematical expressions.

Consider a motor as a system. The input to the motor is a voltage V and the output is the speed of rotation ω of the shaft. For many systems there is a reasonably *linear relationship* between the input and the output. This means that the output is proportional to the input and if the input is doubled the output is doubled, i.e. if the input is multiplied by some constant multiplier then the output is multiplied by the same multiplier. It also means that if input 1 produces output 1 and input 2 produces output 2 then an input equal to the sum of inputs 1 and 2 will produce an output equal to the sum of outputs 1 and 2. Thus if there is a linear relationship between the output and input for the motor, then the mathematical model is

$$\omega = GV$$

where G is the constant of proportionality. This relationship implies that if the voltage is changed then there should be an immediate corresponding change in the shaft speed of rotation. This will not be the case, since the motor will take some time for the shaft to get up to the new speed. The relationship is thus just the relationship between voltage and speed when the system has had time to settle down to any change in input, i.e. it refers to what are called the *steady-state* conditions. Thus, to make this clear, we could write the equation as

Steady-state value of $\omega = G$(steady-state value of V)

Hence

$$G = \frac{\text{steady-state value of } \omega}{\text{steady-state value of } V}$$

The constant G is called the *transfer function* or *gain* of the system. Thus, in general, we can define the transfer function as the ratio of the steady-state output to steady-state input for a system or subsystem.

Transfer function $G = \dfrac{\text{steady-state output}}{\text{steady-state input}}$ [1]

For example, inserting a coin into a chocolate bar machine can lead to the output of one chocolate bar. The transfer function, for the steady-state, is thus 1 bar/coin. Assuming that the system is linear means that for an input of two such coins we would obtain two chocolate bars. A temperature measurement system might have an input of 10°C and produce a steady-state output of 5.0 mV. Such a system has a transfer function of 0.5 mV/°C. Assuming the system is linear enables us to predict that if the input was 20°C then the steady-state output would be 10.0 mV. The mathematical model of the system is thus

Steady-state output in mV = 0.5 × steady-state input in °C

The above is a discussion of a system having a linear relationship between input and output. Real systems can often however exhibit non-linear behaviour. In many cases, however, such systems are linear provided the input signals are kept within certain limits. Thus for the chocolate bar machine the system is linear provided you do not put more coins into the machine than there are chocolate bars. An amplifier might only be linear for input signals up to a certain size.

Later chapters in this book will consider definitions of the transfer function which can take into account not only the steady-state values of the input and output but also the transient changes that take place with time. Thus the remainder of this chapter, in basing the discussions on the behaviour of systems in terms of steady-state values, has thus to be considered as representing only a very simplistic view of control systems.

Example 6

A motor has a transfer function of 500 rev/min per volt. What will be the steady-state output speed for such a motor when the input is 12 V?

Answer

Using equation [1]

$$\text{Transfer function } G = \frac{\text{steady-state output}}{\text{steady-state input}}$$

then

$$\text{Steady-state output} = G(\text{steady-state input})$$
$$= 500 \times 12 = 6000 \text{ rev/min}$$

Mathematical models for open-loop systems

There are many situations where the transfer function is required for a number of elements in series. Consider three elements in series, as in Fig. 1.24. The three elements can be an open-loop system since there is no feedback loop or just three series elements in some bigger system.

For element 1 the transfer function G_1 is the output θ_1 divided by the input θ_i. Thus

$$G_1 = \frac{\theta_1}{\theta_i}$$

Fig. 1.24 The transfer function with an open-loop system

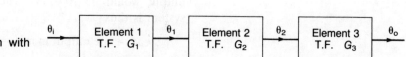

θ_i → | Element 1 T.F. G_1 | → θ_1 → | Element 2 T.F. G_2 | → θ_2 → | Element 3 T.F. G_3 | → θ_o

For element 2 the transfer function G_2 is the output θ_2 divided by its input θ_1. Thus

$$G_2 = \frac{\theta_2}{\theta_1}$$

For element 3 the transfer function G_3 is the output θ_o divided by its input θ_2. Thus

$$G_3 = \frac{\theta_o}{\theta_2}$$

The overall transfer function of the system is the output θ_o divided by the input θ_i. But this can be written as

$$\frac{\theta_o}{\theta_i} = \frac{\theta_1}{\theta_i} \times \frac{\theta_2}{\theta_1} \times \frac{\theta_o}{\theta_2}$$

Hence, for the open-loop system

$$\text{Transfer function} = G_1 \times G_2 \times G_3 \qquad [2]$$

The overall open-loop transfer function is the product of the transfer functions of the individual elements. This applies however many elements there are connected in series.

Example 7

The measurement system used with a control system consists of two elements, a sensor and a signal conditioner in series (Fig. 1.25). If the sensor has a transfer function of 0.1 mA/Pa and the signal conditioner a transfer function of 20, what will be the overall transfer function of the measurement system?

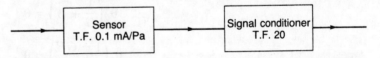

Fig. 1.25 Example 7

Answer

The sensor and the signal conditioner are in series so the combined transfer function of the two elements is the product of the transfer functions of the individual elements.

$$\text{Transfer function} = 0.1 \times 20 = 2\,\text{mA/Pa}$$

Mathematical models for closed-loop systems

Figure 1.26 shows a simple closed-loop system. If θ_i is the reference value, i.e. the input, and θ_o the actual value, i.e. the output, of the system then the transfer function of the entire control system is

$$\text{Transfer function} = \frac{\text{output}}{\text{input}} = \frac{\theta_i}{\theta_o}$$

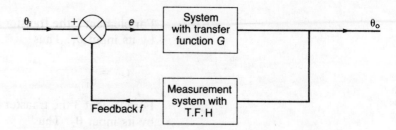

Fig. 1.26 The transfer function with a closed-loop system

Each subsystem within the overall system has its own transfer function. Thus if the system being controlled has a transfer function G then with its input of the error signal e and output of θ_o,

$$G = \frac{\theta_o}{e}$$

If the feedback path has a transfer function H then with its input of θ_o and output f,

$$H = \frac{f}{\theta_o}$$

The error signal e is the difference between θ_i and f, the feedback signal f being a measure of the output of the entire system,

$$e = \theta_i - f$$

Thus substituting for e and f using the above two equations,

$$\frac{\theta_o}{G} = \theta_i - H\theta_o$$

$$\theta_o\left(\frac{1}{G} + H\right) = \theta_i$$

$$\theta_o\left(\frac{1 + GH}{G}\right) = \theta_i$$

Hence the overall transfer function of the closed-loop control system is

$$\text{Transfer function} = \frac{\theta_o}{\theta_i} = \frac{G}{1 + GH} \qquad [3]$$

The above equation is for negative feedback. With positive feedback the denominator of the equation becomes $(1 - GH)$.

With the closed-loop system, G is termed the *forward-path transfer function* since it is the transfer function relating to the signals moving forward through the system from input to output. GH is termed the *loop transfer function* since it is the term that occurs in the expression as a result of there being a feedback loop.

Example 8

A speed-controlled motor has an amplifier-relay-motor system with a combined transfer function of 600 rev/min per volt and a feedback loop measurement system with a transfer function of 3 mV per rev/min, as illustrated in Fig. 1.27. What is the transfer function of the total system?

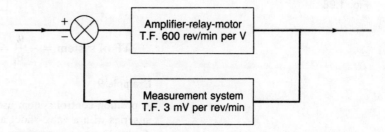

Fig. 1.27 Example 8

Answer

The system will have negative feedback and so the overall transfer function is given by equation [3] as

$$\text{Transfer function} = \frac{G}{1 + GH}$$

$$= \frac{600}{1 + 600 \times 0.003}$$

$$= 214 \text{ rev/min per volt}$$

Mathematical models for multi-element closed-loop systems

Consider the closed-loop system shown in Fig. 1.28. The transfer function for the entire system can be obtained by first determining the transfer function for the three elements in series. Since these have transfer functions G_1, G_2 and G_3, then their combined transfer function is

$$\text{TF of series elements} = G_1 \times G_2 \times G_3$$

The closed-loop system of Fig. 1.28 can now be replaced by a simpler equivalent system, as in Fig. 1.29. It is now just a single element with a transfer function of $G_1 \times G_2 \times G_3$ and a feedback loop with a transfer function H. The overall transfer function for this system is thus

Fig. 1.28 The transfer function with a multi-element closed-loop system

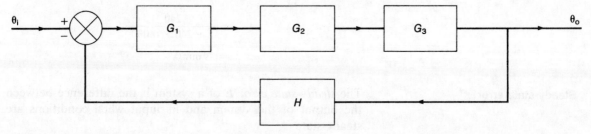

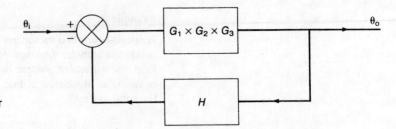

Fig. 1.29 The equivalent system for Fig. 1.28

$$\text{TF of system} = -\frac{\theta_o}{\theta_i} = \frac{G_1 \times G_2 \times G_3}{1 + (G_1 \times G_2 \times G_3)H} \qquad [4]$$

Example 9

A position-control system used with a machine tool has an amplifier in series with a valve-slider arrangement and a feedback loop with a displacement measurement system (Fig. 1.30). If the transfer functions are as follows, what is the overall transfer function for the control system?

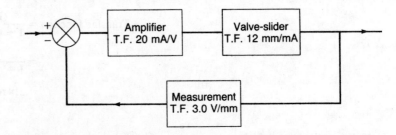

Fig. 1.30 Example 9

Transfer functions: amplifier 20 mA/V, valve-slider arrangement 12 mm/mA, measurement system 3.0 V/mm.

Answer

The amplifier and the valve-slider arrangement are in series so the combined transfer function for the two elements is the product of their separate transfer functions,

TF for the series elements = $20 \times 12 = 240$ mm/V

These elements have a feedback loop with a transfer function of 30 mV/mm. Thus the overall transfer function of the control system is

$$\text{Transfer function} = \frac{G}{1 + GH}$$

$$= \frac{240}{1 + 240 \times 0.030}$$

$$= 29 \text{ mm/V}$$

Steady-state error

The *steady-state error* E of a system is the difference between the output of the system and its input when conditions are steady-state

$$E = \theta_o - \theta_i$$

Since for a system with an overall transfer function of G_s

$$G_s = \frac{\theta_o}{\theta_i}$$

then

$$E = G_s\theta_i - \theta_i = \theta_i(G_s - 1) \qquad [5]$$

For an open-loop system the steady-state error can be written, using equation [2], as

$$E = \theta_i(G_1G_2G_3 - 1) \qquad [6]$$

where G_1, G_2 and G_3 are the transfer functions of the various elements in the system. For the error to be zero then $G_1G_2G_3$ must be equal to 1. While this can be arranged at the setting-up or calibration of such a system, because transfer functions tend to change due to environmental changes it is inevitable that steady-state errors will occur.

For a closed-loop system the steady-state error, equation [5], can be written, using equation [3], as

$$E = \theta_i\left(\frac{G}{1 + GH} - 1\right) \qquad [7]$$

where G is the transfer function of the forward-path elements, i.e. $G = G_1G_2G_3$ for three series elements with transfer functions of G_1, G_2 and G_3, and H is the transfer function of the measurement system. For a zero steady-state error we must have $G = 1 + GH$ so that $G/(1 + GH)$ has the value 1. Like the open-loop closed system, transfer functions will tend to change with environmental conditions. However, if GH is much greater than 1, equation [7] approximates to

$$E = \theta_i\left(\frac{1}{H} - 1\right)$$

and so changes in the transfer functions of the forward-path elements have virtually no effect on the error. Thus the sensitivity of a closed-loop system to such effects is considerably less than that of an open-loop system. This is one of the great advantages that closed-loop systems have over open-loop systems.

Example 10

Figure 1.31 shows a controller with a transfer function of 12 and a motor with a transfer function of 0.10 rev/min per V.
(a) What will be the steady-state error when the system is an open-loop control system and how will the error change if, as a result of environmental changes, the transfer function of the motor changes by 10%?

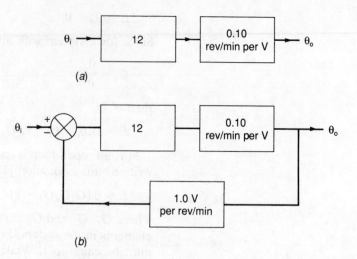

Fig. 1.31 Example 10: (a) open loop, (b) closed loop

(b) What will be the steady-state error when the system is a closed-loop control system with the feedback loop having a transfer function of 1.0 V per rev/min and how will the error change if, as a result of environmental changes, the transfer function of the motor changes by 10%?

Answer

(a) Using equation [6], then before any changes occur

$$E = \theta_i(G_1G_2 - 1)$$

$$E = \theta_i(12 \times 0.10 - 1) = 0.2\,\theta_i$$

If there is a 10% change in the transfer function of the motor, i.e. to 0.11 rev/min per V, then

$$E = \theta_i(12 \times 0.11 - 1) = 0.32\,\theta_i$$

The error has increased by a factor 16.

(b) Using equation [7]

$$E = \theta_i\left(\frac{G_1G_2}{1 + G_1G_2H} - 1\right)$$

then before any changes occur

$$E = \theta_i\left(\frac{12 \times 0.10}{1 + 12 \times 0.10 \times 1.0} - 1\right) = -0.45\,\theta_i$$

If there is a change of 10% in the transfer function of the motor to 0.11 rev/min per V then

$$E = \theta_i\left(\frac{12 \times 0.11}{1 + 12 \times 0.11 \times 1.0} - 1\right) = -0.43\,\theta_i$$

The change in the error is considerably less than the change that occurred with the open-loop system. The closed-loop system has a much lower sensitivity to environmental changes than does the open-loop system.

Example 11

A closed-loop control system has a forward-path transfer function of 10. What should be the transfer function of the feedback path if there is to be zero steady-state error?

Answer

Using equation [7]

$$E = \theta_i \left(\frac{G}{1 + GH} - 1 \right)$$

for E to be zero then

$$\frac{G}{1 + GH} = 1$$

hence with $G = 10$ then

$$10 = 1 + 10H$$

and so H must be 0.9.

Effect of disturbances

An important consideration with a control system is the effect of any disturbances. Thus in a domestic heating system where there is an open-loop control system involving a heater being switched on to obtain the required room temperature, what will happen if someone opens a window and lets a blast of cold air into the room? We can incorporate such a disturbance in the block diagram for the system in the way shown in Fig. 1.32. The disturbance θ_d is added to the process output in this case. For such a situation we thus have

$$\theta_o = G_1 G_2 \theta_i + \theta_d \tag{8}$$

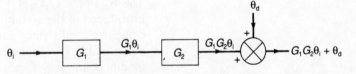

Fig. 1.32 Disturbance with open-loop control system

The θ_d term is the steady-state error added to the system by the presence of the disturbance.

If the disturbance had been added to the system between elements 1 and 2, (as in Fig. 1.33), then

$$\theta_o = (G_1 \theta_i + \theta_d) G_2 = G_1 G_2 \theta_i + G_2 \theta_d \tag{9}$$

The $G_2 \theta_d$ term is the steady-state error added to the system by the presence of the disturbance.

With a closed-loop system, e.g. in a domestic heating system if the thermostat setting is not changed and someone opens a window and lets a blast of cold air into the room, we can represent the control system subject to such a disturbance by a

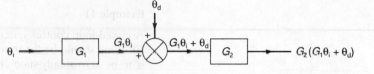

Fig. 1.33 Disturbance with open-loop control system

block diagram of the form shown in Fig. 1.34. The error signal to element 1 is $(\theta_i - f)$, where f is the feedback signal. The signal output from element 1 and so input to element 2 is thus $G_1(\theta_i - f)$. The output from element 2 is thus $G_1G_2(\theta_i - f)$. The disturbance adds to this at this point and so the output θ_o is

$$\theta_o = G_1G_2(\theta_i - f) + \theta_d$$

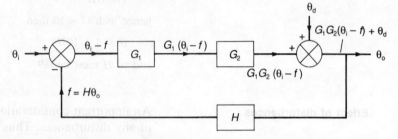

Fig. 1.34 Disturbance with closed-loop control system

But the feedback f is $H\theta_o$. Hence
$$\theta_o = G_1G_2(\theta_i - H\theta_o) + \theta_d$$

Rearranging this gives
$$\theta_o(1 + G_1G_2H) = G_1G_2\theta_i + \theta_d$$
$$\theta_o = \theta_i\left(\frac{G_1G_2}{1 + G_1G_2H}\right) + \theta_d\left(\frac{1}{1 + G_1G_2H}\right) [10]$$

The $\theta_d[1/(1 + G_1G_2H)]$ term is the steady-state error that is introduced to the system by the disturbance. If this equation is compared with the comparable open-loop situation, i.e. equation [8], it will be seen that the effect of the disturbance is modified by the factor $1/(1 + G_1G_2H)$. This property of modifying the effect of a disturbance is called *disturbance rejection*.

Figure 1.35 shows the closed-loop system with the disturbance occurring between the two forward-path elements. For such a system the input to the first element is $(\theta_i - f)$ and its output is thus $G_1(\theta_i - f)$. This is combined with the disturbance θ_d to give the input to the second element. Thus the output from this element θ_o is

$$\theta_o = G_2[G_1(\theta_i - f) + \theta_d]$$

Since $f = H\theta_o$, then
$$\theta_o = G_2[G_1(\theta_i - H\theta_o) + \theta_d]$$

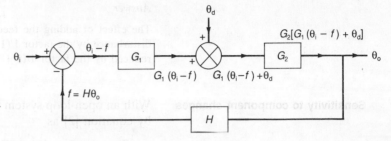

Fig. 1.35 Disturbance with closed-loop control system

This can be simplified to

$$\theta_o[1 + G_1G_2H] = G_2G_1\theta_i + G_2\theta_d$$

$$\theta_o = \theta_i\left(\frac{G_2G_1}{1 + G_1G_2H}\right) + \theta_d\left(\frac{G_2}{1 + G_1G_2H}\right) [11]$$

The $\theta_d[G_2/(1 + G_1G_2H)]$ term is the steady-state error that is introduced to the system by the disturbance. If this equation is compared with the comparable open-loop situation, i.e. equation [9], it will be seen that the effect of the disturbance is modified in the closed-loop situation by the factor $G_2/(1 + G_1G_2H)$ and in the open-loop situation by just G_2. The factor $1/(1 + G_1G_2H)$ is thus a measure of how much the effects of the disturbance are modified by the feedback loop. This property of modifying the effect of a disturbance is called disturbance rejection.

Thus wherever in the closed-loop system a disturbance occurs, its effect is reduced by the factor $1/(1 + GH)$, where G is the forward-path transfer function ($G = G_1G_2$) and H is the feedback transfer function. This is one of the advantages closed-loop control systems have over open-loop control systems: they are much better at alleviating the effects of disturbances in the system.

Disturbances can arise in a number of ways, the term disturbance being taken to mean any unwanted signal which affects the output of the system. Disturbances can arise from signals from extraneous sources, e.g. someone opening a window and so affecting the heating system for the room or perhaps the wind affecting the steering of a radar dish or perhaps a pot-hole in the road affecting the steering of a motor car along the road. Disturbances can also arise within the system, e.g. electrical noise in an amplifier.

Example 12

An electronic amplifier has a transfer function of 100. How much better will the amplifier be at rejecting internally generated noise if the amplifier is given a feedback loop with a transfer function of 10?

Answer

The effect of adding the feedback loop is to reduce the effect of a disturbance by the factor $1/(1 + GH)$. Thus the effect of the noise is reduced by the factor $1/(1 + 100 \times 10) = 1/1001$.

Sensitivity to component changes

With an open-loop system the overall transfer function is given by equation [2] as

$$\text{Transfer function} = G_1 \times G_2 \times G_3$$

where G_1, G_2 and G_3 are the transfer functions of the elements in the system. Changes in the characteristics of such elements with time and environmental conditions may result in a change in transfer function. Thus, for example, if one element is a motor then an increase in friction at the bearings could result in a decrease in the transfer function for the motor. A change in transfer function of element 1 of ΔG_1 means a change in the overall transfer function for the open-loop system of

$$\text{Change in transfer function} = \Delta G_1 \times G_2 \times G_3 \qquad [12]$$

With a closed-loop system the overall transfer function is given by equation [4] as

$$\text{Transfer function} = \frac{G_1 G_2 G_3}{1 + G_1 G_2 G_3 H}$$

where G_1, G_2 and G_3 are the transfer functions of the elements in the forward path and H the transfer function of the feedback path. If $G_1 G_2 G_3 H$ is much greater than 1 then the expression approximates to

$$\text{Transfer function} = \frac{1}{H}$$

Any change in the transfer function of, say, element 1 in the forward path, will have negligible effect on the overall transfer function. This is in stark contrast to the situation with an open-loop system and this insensitivity to changes in the characteristics of forward-path elements is one of the advantages of closed-loop systems over open-loop systems. However, a change in the transfer function of the feedback path will produce a corresponding change in the overall transfer function of the system. The closed-loop system is not insensitive to changes in the characteristics of elements in the feedback path.

Example 13

An operational amplifier has an open-loop transfer function of 200 000 and a feedback path with a transfer function of 0.1. What will be the percentage change in the overall transfer function if the transfer function of the amplifier changes by 10%?

Answer

For the closed-loop system, using equation [3],

$$\text{Transfer function} = \frac{G}{1 + GH}$$

Hence, initially,

$$\text{Transfer function} = \frac{200\,000}{1 + 200\,000 \times 0.1} = 9.9995$$

The change in transfer function could mean a transfer function of 180 000 for the amplifier. Hence the new transfer function is given by

$$\text{Transfer function} = \frac{180\,000}{1 + 180\,000 \times 0.1} = 9.9994$$

The change in the overall transfer function is 0.0001 and hence a percentage change of 0.01%. Thus used with no feedback path the amplifier would have had a 10% change in transfer function; with the feedback path it is 0.01%.

Stability of control systems

In mechanical terms an object can be said to be in stable equilibrium if when you give it a push it returns back to its original position when you stop pushing. An example of this is a ball resting in a spherical dish (Fig. 1.36). When the ball is pushed it moves up the side of the dish, but when you stop pushing the ball it soon returns back to its position at rest in the centre of the dish. However, the position would be unstable if the sphere had been resting on the outside of an upturned dish, any slight push would cause the sphere to roll off and it would not return to its original position when the pushing stopped.

Returns when   Pushed and does
push stops   not return

Fig. 1.36 (a) Stable, (b) unstable

In general, a system is said to be *stable* if when it is subject to a bounded input or disturbance then the output is bounded. A bounded input or output is one that has a finite size. Thus in the case of the ball, the input is initially zero followed by a push which does not continue indefinitely but ceases after a while. The output in the stable condition is that the push causes the the ball to move and become displaced from its initial rest position, but eventually the ball ceases moving and the displacement does not continue increasing or changing. In the unstable condition, the output of the displacement goes on increasing, i.e. a finite size input can produce an output which is without limit.

The condition for stability can also be expressed as that a system is stable if subject to an impulse the output returns eventually to zero.

Open-loop control systems are inherently stable. A finite input causes an output which is finite and does not go on changing indefinitely with time. Increasing the transfer function of an element in such a system has no effect on the stability of the system. Thus for an open-loop system called a chocolate bar machine, putting in a suitable coin will result in the output of a chocolate bar, not a supply of chocolate bars which continues indefinitely. Changing the rate of exchange, the cost of each bar, has no effect on the stability of the system.

Closed-loop systems can however show instability. Such instability can occur as a result of time delays occurring between the change in the variable and the feedback signal resulting in the system responding. An example of this could be a control system for a robot hand, or the hand of a drunk, picking up a cup of coffee. The hand is moved towards the cup of coffee and suppose it veers off to the right of the required path. This deviation is noticed and a signal sent to the hand to start moving to the left. This signal is continued in the ideal situation until the hand reaches the required path. However, if there are time delays in the system the action may continue for too long and the hand may overshoot the required path before the system reacts. The hand is then too far to the left. A signal is then sent to the hand to move to the right. Again because of delays the hand overshoots the required path and ends up now to the right of the required position. The entire cycle can go on repeating with the hand staggering back-and-forth, to the right and left, of the required path. Thus an initial disturbance can produce unstable condition. The instability depends on the forward-path transfer function since a large value for this results in a greater amount of movement of the hand during the delay time.

Another example of instability can occur when a person takes a shower and manually adjusts the temperature of the water by means of a mixer tap which enables the relative amounts of cold and hot water to be determined. Suppose initially the water is too cold. The hot water element is then increased. However, there is a time delay before the hot water reaches the shower head. Thus if the person does not wait but just responds to the water temperature he/she will continue to increase the hot water element. The result will be that when the hotter water reaches the shower head it will rapidly run too hot. The person then increases the cold water element. Again because of a time delay some time will elapse before the water at the shower head cools down. So he/she continues to increase the cold water element. The result will be that when the colder water reaches the shower head it will rapidly run too cold. So the cycle is repeated with the result that the water temperature at the shower head oscillates widely.

The above is just a very sketchy consideration of stability, a more detailed discussion appears in Chapter 6.

Closed-path versus open-path control

The advantages of having a feedback path and hence a closed-loop system rather than an open-loop system can be summarized as:

1 More accurate in matching actual to required values for the variable.
2 Less sensitive to disturbances.
3 Less sensitive to changes in component characteristics.
4 Increased speed of response and hence bandwidth, i.e. range of frequencies over which the system will respond.

But there are the disadvantages:

1 There is a loss in gain in that the transfer function of an open-loop system is reduced from G to $G/(1 + GH)$ by a feedback loop with a transfer function H.
2 There is a greater chance of instability.
3 The system is more complex and so, not only more expensive, but more prone to breakdown.

Problems

1 Explain the difference between open-loop and closed-loop control systems.
2 State which of the following are open-loop and which closed-loop control systems and give reasons for your statements:
    (*a*) An electric kettle which switches off when the water boils.
    (*b*) A refrigerator.
    (*c*) An electric hotplate with no thermostat.
3 Traffic lights at a road crossing can be open-loop or closed-loop control systems. Explain how the systems would differ.
4 Draw box diagrams showing the subsystems in the following closed-loop systems:
    (*a*) An automatic (exposure) camera.
    (*b*) A thermostatically controlled oven.
    (*c*) An automatic light which comes on when it gets dark and goes off when it becomes light.
5 Explain the difference between two-step and proportional control strategies.
6 What type of control strategy is likely to be being exercised in the following control systems:

(a) A domestic refrigerator?

(b) A bread toaster?

(c) Picking up a cup of coffee?

7   The automatic control system for the temperature of a bath of liquid consists of a reference voltage fed into a differential amplifier. This is connected to a relay which then switches on or off the electrical power to a heater in the liquid. Negative feedback is provided by a measurement system which feeds a voltage into the differential amplifier. Sketch a block diagram of the system and explain how the error signal is produced.

8   A temperature-measurement system has a thermometer which produces a resistance change of $0.007\,\Omega/°C$ connected to a Wheatstone bridge which produces a current change of $20\,mA/\Omega$. What is the overall transfer function of the system?

9   Explain what is meant by there being a linear relationship between the input and output of a system.

10   Figure 1.37 shows a system used to control the rate of flow of liquid along a pipe.

(a) Explain how the system operates.

(b) What will be the transfer function for the feedback loop if the flow meter has a transfer function of 2 kPa per m/s and the pressure to current converter 1.0 mA per kPa?

(c) What will be the transfer function for the forward path if the current to a pressure converter has a transfer function of 6 kPa per mA and the control valve 0.1 m/s per kPa?

(d) What will be the overall transfer function of the control system?

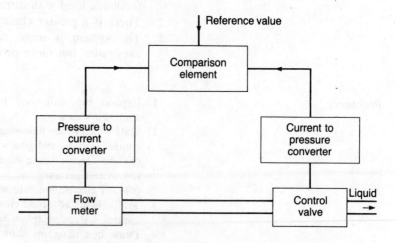

Fig. 1.37 Problem 10

11   What will be the steady-state error for an open-loop temperature-control system consisting of a controller with a transfer function of 1.0 in series with a heater with a transfer function of 0.80 °C/V and what will be the percentage change in the steady-state error if the transfer function of the heater decreases by 1%?

12   What will be the steady-state error for a closed-loop temperature control system consisting of a controller with a transfer function of 20 in series with a heater with a transfer function of 0.80 °C/V

and a feedback loop with a transfer function of 10 V/°C, and what will be the percentage change in the steady-state error if the transfer function of the heater decreases by 1%?

13 Explain why closed-loop feedback systems are much better at rejecting disturbances than open-loop systems.

14 An open-loop system has a transfer function of K. What will be the effect on the output of that system if the transfer function is reduced to $\frac{1}{2}K$? What would be the effect if the system had a feedback loop with a transfer function of 1?

15 What are the advantages and disadvantages of an amplifier being given a feedback loop?

2 System models

Introduction

In order to analyse control systems *mathematical models* are needed of the elements used in such systems. These are equations representing the relationship between the input and output of a system (see Chapter 1). The basis for any mathematical model is provided by the fundamental physical laws governing the behaviour of an element. In this chapter a range of systems will be considered, including mechanical, electrical, thermal and fluid examples.

Like a child building houses, cars, cranes, etc., from a number of basic building blocks, systems can be made up from a range of building blocks. Each building block is considered to have a single property or function. Thus, to take a simple example, an electrical circuit system may be made up from building blocks which represent the behaviour of resistors, capacitors and inductors. The resistor building block is assumed to have purely the property of resistance, the capacitor purely that of capacitance and the inductor purely that of inductance. By combining these building blocks in different ways a variety of electrical circuit systems can be built up and the overall input-output relationships obtained for the system by combining in an appropriate way the relationships for the building blocks. Thus a mathematical model for the system can be obtained. A system built up in this way is called a *lumped parameter* system. This is because each parameter, i.e. property or function, is considered independently.

Since there are similarities in behaviour of building blocks used in mechanical, electrical, thermal and fluid systems we do not even need different forms of 'mathematical building blocks' for the different types of system. This chapter is about the basic building blocks and their combination to produce mathematical models for physical, real, systems.

Mechanical system building blocks

The basic forms of the mechanical system building blocks are springs, dashpots and masses. Springs represent the stiffness of a system, dashpots the forces opposing motion, i.e. frictional or damping effects, and masses the inertia or resistance to acceleration. All these building blocks can be considered to have a force as an input and a displacement as an output.

The stiffness of a *spring* is described by the relationship between the forces F used to extend or compress a spring and the resulting extension or compression x (Fig. 2.1). In the case of a spring where the extension or compression is proportional to the applied forces, i.e. a linear spring,

$$F = kx \qquad [1]$$

where k is a constant. The bigger the value of k the greater the forces have to be to stretch or compress the spring and so the greater the stiffness. The object applying the force to stretch the spring is also acted on by a force, the force being that exerted by the stretched spring (Newton's third law). This force will be in the opposite direction and equal in size to the force used to stretch the spring, i.e. kx.

The *dashpot* building block represents the type of forces experienced when we endeavour to push an object through a fluid or move an object against frictional forces. The faster the object is pushed the greater becomes the opposing forces. The dashpot which is used pictorially to represent these damping forces which slow down moving objects consists of a piston moving in a closed cylinder (Fig. 2.2). Movement of the piston requires the fluid on one side of the piston to flow through or past the piston. This flow produces a resistive force. In the ideal case, the damping or resistive force F is proportional to the velocity v of the piston. Thus

$$F = cv$$

where c is a constant. The larger the value of c the greater is the damping force at a particular velocity. Since velocity is the rate of change of displacement x of the piston, i.e. $v = dx/dt$, then

$$F = c\frac{dx}{dt} \qquad [2]$$

Thus the relationship between the displacement x of the piston, i.e. the output, and the force as the input is a relationship depending on the rate of change of the output.

The *mass* building block (Fig. 2.3) exhibits the property that the bigger the mass the greater is the force required to give it a specific acceleration. The relationship between the force F and the acceleration a is (Newton's second law)

$$F = ma$$

where the constant of proportionality between the force and

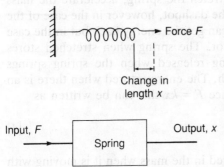

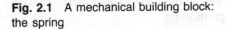

Fig. 2.1 A mechanical building block: the spring

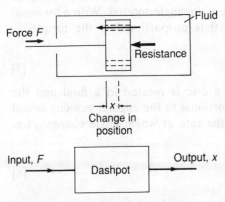

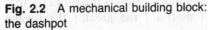

Fig. 2.2 A mechanical building block: the dashpot

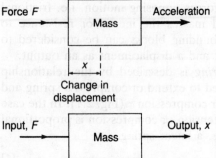

Fig. 2.3 A mechanical building block: mass

the acceleration is the constant called the mass m. Acceleration is the rate of change of velocity, i.e. dv/dt, and velocity v is the rate of change of displacement x, i.e. $v = dx/dt$. Thus

$$F = ma = m\frac{dv}{dt} = m\frac{d(dx/dt)}{dt} = m\frac{d^2x}{dt^2} \qquad [3]$$

Energy is needed to stretch the spring, accelerate the mass and move the piston in the dashpot, however in the case of the spring and the mass we can get the energy back but in the case of the dashpot we cannot. The spring when stretched stores energy, the energy being released when the spring springs back to its original length. The energy stored when there is an extension x is $\frac{1}{2}kx^2$. Since $F = kx$ this can be written as

$$E = \frac{1}{2}\frac{F^2}{k} \qquad [4]$$

There is also energy stored in the mass when it is moving with a velocity v, the energy being referred to as kinetic energy, and released when it stops moving.

$$E = \frac{1}{2}mv^2 \qquad [5]$$

However, there is no energy stored in the dashpot. It does not return to its original position when there is no force input. The dashpot dissipates energy rather than storing it, the power P dissipated depending on the velocity v and being given by

$$P = cv^2 \qquad [6]$$

The spring, dashpot and mass are the building blocks for mechanical systems where forces and straight-line displacements are involved without any rotation. If there is rotation then the equivalent three building blocks are a torsional spring, a rotary damper and the moment of inertia, i.e. the inertia of a rotating mass. With such building blocks the inputs are torque and the output is the angle rotated. With a *torsional spring* the angle rotated θ is proportional to the torque T. Hence

$$T = k\theta \qquad [7]$$

With the *rotary damper* a disc is rotated in a fluid and the resistive torque T is proportional to the angular velocity ω and since angular velocity is the rate at which angle changes, i.e. $d\theta/dt$,

$$T = c\omega = c\frac{d\theta}{dt} \qquad [8]$$

The *moment of inertia* building block exhibits the property that the greater the moment of inertia I the greater the torque needed to produce an angular acceleration α.

$$T = I\alpha$$

Thus, since angular acceleration is the rate of change of angular velocity, i.e. $d\omega/dt$, and angular velocity is the rate of change of angular displacement, then

$$T = I\frac{d\omega}{dt} = I\frac{d(d\theta/dt)}{dt} = I\frac{d^2\theta}{dt^2} \qquad [9]$$

The torsional spring and the rotating mass store energy, the rotary damper just dissipates energy. The energy stored by a torsional spring when twisted through an angle θ is $\frac{1}{2}k\theta^2$ and since $T = k\theta$ this can be written as

$$E = \frac{1}{2}\frac{T^2}{k} \qquad [10]$$

The energy stored by a mass rotating with an angular velocity ω is the kinetic energy E, where

$$E = \frac{1}{2}I\omega^2 \qquad [11]$$

The power P dissipated by the rotatory damper when rotating with an angular velocity ω is

$$P = c\omega^2 \qquad [12]$$

Table 2.1 summarizes the equations defining the characteristics of the mechanical building blocks when there is, in the

Table 2.1 Characteristics of mechanical building blocks

Building block	Describing equation	Energy stored/ power dissipated
Energy storage		
Translational spring	$F = kx$	$E = \dfrac{1}{2}\dfrac{F^2}{k}$
Torsional spring	$T = k\theta$	$E = \dfrac{1}{2}\dfrac{T^2}{k}$
Mass	$F = m\dfrac{d^2x}{dt^2}$	$E = \frac{1}{2}mv^2$
Moment of inertia	$T = I\dfrac{d^2\theta}{dt^2}$	$E = \frac{1}{2}I\omega^2$
Energy dissipation		
Translational dashpot	$F = c\dfrac{dx}{dt}$	$P = cv^2$
Rotational damper	$T = c\dfrac{d\theta}{dt}$	$P = c\omega^2$

case of straight line displacements (termed translational), a force input F and a displacement x output and in the case of rotation a torque T and angular displacement θ.

Building up a mechanical system

Many systems can be considered to be essentially a mass, a spring and dashpot combined in the way shown in Fig. 2.4. To evaluate the relationship between the force and displacement for the system the procedure to be adopted is to consider just one mass, and just the forces acting on that body. A diagram of the mass and the forces acting on it is called a *free-body diagram*. When several forces act on a body simultaneously, their single equivalent resultant can be found by vector addition. If the forces are all acting along the same line or parallel lines, this means that the resultant or net force acting on the block is the algebraic sum. Thus for the mass in Fig. 2.4, if we consider just the forces acting on that block then the net force applied to the mass is the applied force F minus the force resulting from the stretching or compressing of the spring and minus the force from the damper. Thus, using equations [1] and [2]

Net force applied to mass $m = F - kx - cv$

$$= F - kx - c\frac{\mathrm{d}x}{\mathrm{d}t}$$

where v is the velocity with which the piston in the dashpot and hence the mass is moving. This net force is the force applied to the mass to cause it to accelerate. Thus, using equation [3]

Net force applied to mass $= ma = m\dfrac{\mathrm{d}^2x}{\mathrm{d}t^2}$

Hence

$$F - kx - c\frac{\mathrm{d}x}{\mathrm{d}t} = m\frac{\mathrm{d}^2x}{\mathrm{d}t^2}$$

$$m\frac{\mathrm{d}^2x}{\mathrm{d}t^2} + c\frac{\mathrm{d}x}{\mathrm{d}t} + kx = F \qquad [13]$$

This equation, called a differential equation, describes the relationship between the input of force F to the system and the output of displacement x.

This equation is usually written in a different form, the constants m, c and k being replaced by other constants for the system. In the absence of the damping a mass m on the end of

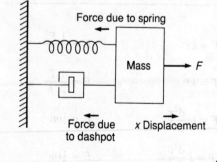

Force due to spring

Mass

$\longrightarrow F$

Force due to dashpot

x Displacement

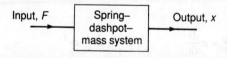

Input, F → Spring–dashpot–mass system → Output, x

Fig. 2.4 Spring-dashpot-mass system

a spring would freely oscillate with a *natural angular frequency* ω_n given by

$$\omega_n = \sqrt{(k/m)}$$

The motion is however damped, a *damping ratio* ζ being used to define the extent of the damping. This damping ratio (see Chapter 3 for an explanation of this relationship) is given by

$$\zeta = \frac{c}{2\sqrt{(mk)}}$$

Thus the equation becomes

$$\frac{1}{\omega_n^2}\frac{d^2x}{dt^2} + \frac{2\zeta}{\omega_n}\frac{dx}{dt} + x = \frac{F}{k} \qquad [14]$$

There are many systems that can be built up from combinations of the spring, dashpot and mass building blocks. Figure 2.5(*a*) shows the model for a machine mounted on the ground and is a basis for studying the effects of ground disturbances on the displacements of the machine bed. Figure 2.5(*b*) shows a model for the wheel and its suspension for a car or truck and can be used for the study of the behaviour that could be expected of the vehicle when driven over a rough road and hence as a basis for the design of the vehicle suspension. The procedure to be adopted for the analysis of such models is just the same as outlined above for the simple spring-dashpot-mass model. A free-body diagram is drawn for each mass in the system, such diagrams showing each mass independently and just the forces acting on it. Then for each mass the resultant of the forces acting on it is then equated to the product of the mass and the acceleration of the mass.

Similar models can be constructed for rotating systems. To evaluate the relationship between the torque and angular displacement for the system the procedure to be adopted is to consider just one rotational mass block, and just the torques acting on that body. When several torques act on a body simultaneously, their single equivalent resultant can be found by addition in which the direction of the torques is taken into account. Thus a system involving a torque being used to rotate a mass on the end of a shaft (Fig. 2.6(*a*)) can be considered to be represented by the rotational building blocks shown in Fig. 2.6(*b*). This is a comparable situation to that analysed above (Fig. 2.4) for linear displacements and yields equations similar to those given in [7] and [8], namely

$$I\frac{d^2\theta}{dt^2} + c\frac{d\theta}{dt} + k\theta = T \qquad [15]$$

$$\frac{1}{\omega_n^2}\frac{d^2\theta}{dt^2} + \frac{2\zeta}{\omega_n}\frac{d\theta}{dt} + \theta = \frac{T}{k} \qquad [16]$$

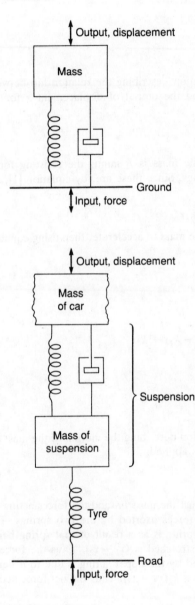

(a)

(b)

Fig. 2.5 Examples of mechanical models: (*a*) a machine mounted on the ground, (*b*) the wheel of a car or truck moving along the road

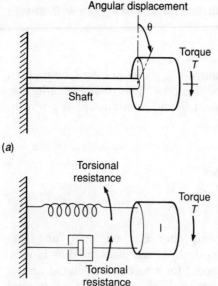

Angular displacement

θ

Torque
T

Shaft

(a)

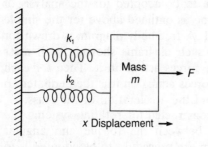

Torsional
resistance

Torque
T

I

Torsional
resistance

(b)

Fig. 2.6 (a) Rotating a mass on the end of a shaft, (b) the building block model

k_1

Mass
m

→ F

k_2

x Displacement →

Fig. 2.7 Example 1

where ω_n is the natural angular frequency for angular rotations and ζ the damping ratio for angular motion with

$$\omega_n = \sqrt{(k/I)}$$

and

$$\zeta = \frac{c}{2\sqrt{(Ik)}}$$

Example 1

Derive the differential equation describing the relationship between the input of the force F and the output of displacement x for the system shown in Fig. 2.7.

Answer

The net force applied to the mass is F minus the resisting forces exerted by each of the springs. Since these are, by equation [1], $k_1 x$ and $k_2 x$, then

Net force $= F - k_1 x - k_2 x$

Since the net force causes the mass to accelerate, then using equation [3]

$$\text{Net force} = m\frac{d^2 x}{dt^2}$$

Hence

$$m\frac{d^2 x}{dt^2} = F - k_1 x - k_2 x$$

$$m\frac{d^2 x}{dt^2} + (k_1 + k_2)x = F$$

Example 2

Derive the differential equation describing the motion of the mass m_1 in Fig. 2.8 when a force F is applied.

Answer

The first step is to consider just the mass m_1 and the forces acting on it (Fig. 2.9). These are the forces exerted by the two springs. The force exerted by the lower spring is as a result of that spring being stretched. The amount it is stretched is $(x_1 - x_2)$. Thus the force is $k_1(x_1 - x_2)$. The force exerted by the upper spring is due to it being stretched by $(x_2 - x_3)$ and so it $k_2(x_3 - x_2)$. Thus the net force acting on the mass is

Net force $= k_1(x_2 - x_1) - k_2(x_3 - x_2)$

This net force will cause the mass to have an acceleration. Hence

$$m\frac{d^2 x}{dt} = k_1(x_2 - x_1) - k_2(x_3 - x_2)$$

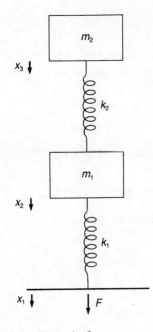

Fig. 2.8 Example 2

But the force causing the extension of the lower spring is F. Thus

$$F = k_1(x_2 - x_1)$$

Hence the equation can be written as

$$m\frac{d^2x}{dt} + k_2(x_3 - x_2) = F$$

Example 3

A motor is used to rotate a load. Devise a model and obtain the differential equation for it.

Answer

This is essentially the situation described by Fig. 2.6(*a*) and thus the model is that described by Fig. 2.6(*b*). The differential equation is thus equation [9], i.e.

$$I\frac{d^2\theta}{dt^2} + c\frac{d\theta}{dt} + k\theta = T$$

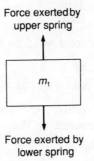

Force exerted by upper spring

m_1

Force exerted by lower spring

Fig. 2.9 Example 2

Electrical system building blocks

The basic building blocks of passive electrical systems are inductors, capacitors and resistors. For an *inductor* the potential difference v across it at any instant depends on the rate of change of current (di/dt) through it.

$$v = L\frac{di}{dt} \qquad [17]$$

where L is the inductance. The direction of the potential difference is in the opposite direction to the potential difference used to drive the current through the inductor, hence the term back e.m.f. The equation can be rearranged to give

$$i = \frac{1}{L}\int v\,dt \qquad [18]$$

For a *capacitor*, the potential difference across it depends on the charge q on the capacitor plates at the instant concerned.

$$v = \frac{q}{C} \tag{19}$$

where C is the capacitance. Since the current i to or from the capacitor is the rate at which charge moves to or from the capacitor plates, i.e.

$$i = \frac{\mathrm{d}q}{\mathrm{d}t}$$

then the total charge q on the plates is given by

$$q = \int i \, \mathrm{d}t$$

and so equation [19] can be written as

$$v = \frac{1}{C} \int i \, \mathrm{d}t \tag{20}$$

Alternatively, since $v = q/C$ then

$$\frac{\mathrm{d}v}{\mathrm{d}t} = \frac{1}{C} \frac{\mathrm{d}q}{\mathrm{d}t}$$

and so, since $i = \mathrm{d}q/\mathrm{d}t$, then

$$i = C \frac{\mathrm{d}v}{\mathrm{d}t} \tag{21}$$

For a *resistor*, the potential difference v across it at any instant depends on the current i through it.

$$v = Ri \tag{22}$$

where R is the resistance.

Both the inductor and capacitor store energy which can then be released at a later time. A resistor does not store energy but just dissipates it. The energy stored by an inductor when there is a current i is

$$E = \tfrac{1}{2}Li^2 \tag{23}$$

The energy stored by a capacitor when there is a potential difference v across it is

$$E = \tfrac{1}{2}Cv^2 \tag{24}$$

The power P dissipated by a resistor when there is a potential difference v across it is

$$P = \frac{1}{R} v^2 \tag{25}$$

Table 2.2 summarizes the equations defining the characteristics of the electrical building blocks when (*a*) the input is current and the output is potential difference and (*b*) when the input is potential difference and the output is current.

Table 2.2 Characteristics of electrical building blocks

Building block	Describing equation (a)	(b)	Energy stored/ power dissipated
Energy storage			
Inductor	$v = L\dfrac{di}{dt}$	$i = \dfrac{1}{L}\displaystyle\int v\,dt$	$E = \tfrac{1}{2}Li^2$
Capacitor	$v = \dfrac{1}{C}\displaystyle\int i\,dt$	$i = C\dfrac{dv}{dt}$	$E = \tfrac{1}{2}Cv^2$
Energy dissipation			
Resistor	$v = Ri$	$i = \dfrac{v}{R}$	$P = \dfrac{1}{R}v^2$

Building up a model for an electrical system

The equations describing how the electrical building blocks can be combined are *Kirchoff's laws*. These can be expressed as:

(a) *Law 1*. The total current flowing towards a junction is equal to the total current flowing from that junction, i.e. the algebraic sum of the currents at the junction is zero.

(b) *Law 2*. In a closed circuit or loop, the algebraic sum of the potential differences across each part of the circuit is equal to the applied e.m.f.

A convenient way of using law 1 is called *node analysis* since the law is applied to each principal node of a circuit, a node being a point of connection or junction between building blocks or circuit elements and a principal node being one where three or more branches of the circuit meet. A convenient way of using law 2 is called *mesh analysis* since the law is applied to each mesh, a mesh being a closed path or loop which contains no other loop.

To illustrate the use of these two methods of analysis to generate relationships, consider the circuit shown in Fig. 2.10. All the components are resistors for this illustrative example. With node analysis a principal node, point A on the figure, is picked and the voltage at the node given a value v_A with reference to some other principal node that has been picked as the reference. In this case it is convenient to pick node B as the reference. We then consider all the currents entering and leaving node A since, according the Kirchoff's first law,

$$i_1 = i_2 + i_3$$

The current entering through R_1 is i_1 and since the potential difference across R_1 is $(v - v_A)$, then

$$i_1R_1 = v - v_A$$

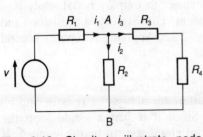

Fig. 2.10 Circuit to illustrate node analysis

The current through R_2 is i_2 and since the potential difference across R_2 is v_A then

$$i_2R_2 = v_A$$

The current i_3 passes through R_3 in series with R_4 and there is a potential difference of v_A across the combination. Hence

$$i_3(R_3 + R_4) = v_A$$

Thus, equating the currents, gives

$$\frac{v - v_A}{R_1} = \frac{v_A}{R_2} + \frac{v_A}{R_3 + R_4}$$

To illustrate the use of mesh analysis for the circuit in Fig. 2.10 it is often convenient to assume there are currents circulating in each mesh in the way shown in Fig. 2.11. Then Kirchoff's second law is applied to each mesh. Thus for the mesh with current i_1 circulating, since the current through R_1 is i_1 and that through R_2 is $(i_1 - i_2)$, then

$$v = i_1R_1 + (i_1 - i_2)R_2$$

$$v = i_1(R_1 + R_2) - i_2R_2$$

Similarly, for the mesh with current i_2 circulating, since there is no source of e.m.f., then

$$0 = i_2R_3 + i_2R_4 + (i_2 - i_1)R_2$$

Rearrangement of this equation gives

$$i_2(R_3 + R_4 + R_2) = i_1R_2$$

Hence substitution for i_2, using this equation, into the equation for the first mesh gives

$$v = i_1(R_1 + R_2) - \frac{i_1R_2^2}{R_3 + R_4 + R_2}$$

$$v = \frac{i_1(R_1R_3 + R_1R_4 + R_1R_2 + R_2R_3 + R_2R_4)}{R_3 + R_4 + R_2}$$

In general, when the number of nodes in a circuit is less than the number of meshes it is easier to employ nodal analysis.

A simple electrical system consists of a resistor and capacitor in series, as in Fig. 2.12. Applying Kirchoff's second law to the circuit loop gives

$$v = v_R + v_C$$

where v_R is the potential difference across the resistor and v_C that across the capacitor. Since it is just a single loop the current through all the circuit elements will be the same, i. If the output from the circuit is the potential difference across

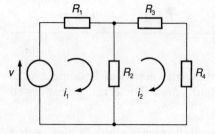

Fig. 2.11 Circuit to illustrate mesh analysis

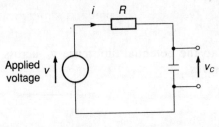

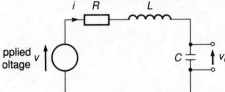

Fig. 2.12 Resistor-capacitor system

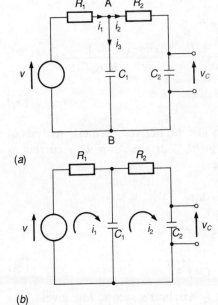

(a)

(b)

Fig. 2.14 (a) nodal analysis,
(b) mesh analysis

Fig. 2.13 Resistor-inductor-capacitor
system

the capacitor, v_C, then since, according to equations [21] and [22], $v_R = iR$ and $i = C(dv_C/dt)$,

$$v = iR + v_C$$

$$v = RC\frac{dv_C}{dt} + v_C \qquad [26]$$

This gives the relationship between the output v_C and the input v.

Figure 2.13 shows a resistor-inductor-capacitor system. If Kirchoff's second law is applied to this circuit loop

$$v = v_R + v_L + v_C$$

where v_R is the potential difference across the resistor, v_L that across the inductor and v_C that across the capacitor. Since there is just a single loop the current i will be the same through all circuit elements. If the output from the circuit is the potential difference across the capacitor v_C, then since, according to equations [21] and [17], $v_R = iR$ and $v_L = L(di/dt)$

$$v = iR + L\frac{di}{dt} + v_C$$

But according to equation [22]

$$i = C\frac{dv_C}{dt}$$

Thus

$$\frac{di}{dt} = C\frac{d(dv_C/dt)}{dt} = C\frac{d^2v_C}{dt^2}$$

Hence

$$v = RC\frac{dv_C}{dt} + LC\frac{d^2v_C}{dt^2} + v_C \qquad [27]$$

Figure 2.14 shows an electrical circuit containing two loops. We first will apply nodal analysis to the problem of determining the relationship between the potential difference v_C across the capacitor and the input voltage v. Point A in Fig. 2.14(a) is taken as a node with a potential v_A relative to node B. For node A

$$i_1 = i_2 + i_3 \qquad [28]$$

The potential difference across R_1 is $(v - v_A)$, thus

$$i_1 R_1 = v - v_A$$

The potential difference across C_1 is v_A, hence equation [21] gives

$$i_3 = C_1 \frac{dv_A}{dt}$$

The current i_2 gives rise to the potential difference v_C across C_2, hence equation [21] gives

$$i_2 = C_2 \frac{dv_C}{dt}$$

Hence the condition for the currents at node A, equation [28], becomes

$$\frac{v - v_A}{R_1} = C_2 \frac{dv_C}{dt} + C_1 \frac{dv_A}{dt} \tag{29}$$

The potential difference across the series combination of R_2 and C_2 is v_A, hence

$$v_A = i_2 R_2 + v_C$$

$$v_A = R_2 C_2 \frac{dv_C}{dt} + v_C$$

Differentiating this gives

$$\frac{dv_A}{dt} = R_2 C_2 \frac{d^2 v_C}{dt^2} + \frac{dv_C}{dt}$$

Hence substituting for v_A and dv_A/dt in equation [29] gives

$$\frac{v}{R_1} - \frac{R_2 C_2}{R_1} \frac{dv_C}{dt} - \frac{v_C}{R_1} = C_2 \frac{dv_C}{dt} + R_2 C_2 C_1 \frac{d^2 v_C}{dt^2} + C_1 \frac{dv_C}{dt}$$

Hence

$$\frac{d^2 v_C}{dt^2} + \left(\frac{R_1 C_1 + R_1 C_2 + R_2 C_2}{R_1 R_2 C_1 C_2} \right) \frac{dv_C}{dt} + \frac{1}{R_1 R_2 C_1 C_2} v_C$$

$$= \frac{v}{R_1 R_2 C_1 C_2} \tag{30}$$

The above equation can also be derived by mesh analysis of the circuit, as in Fig. 2.14(*b*). For the loop with current i_1 Kirchoff's second law gives

$$v = i_1 R_1 + \frac{1}{C_1} \int (i_1 - i_2) dt$$

This can be rearranged to give

$$v = i_1 R_1 + \frac{1}{C_1} \int i_1 dt - \frac{1}{C_2} \int i_2 dt \tag{31}$$

For the loop with current i_2 Kirchoff's second law gives

$$0 = i_2 R_2 + v_C + \frac{1}{C_1} \int (i_2 - i_1) dt$$

This can be rearranged to give

$$0 = i_2 R_2 + v_C + \frac{1}{C_1}\int i_2 dt - \frac{1}{C_1}\int i_1 dt \qquad [32]$$

Since

$$i_2 = C_2 \frac{dv_C}{dt}$$

and

$$v_C = \frac{1}{C_2}\int i_2 dt$$

then equation [32] can be written as

$$0 = R_2 C_2 \frac{dv_C}{dt} + v_C + \frac{C_2 v_C}{C_1} - \frac{1}{C_1}\int i_1 dt$$

Differentiating this equation gives

$$0 = R_2 C_2 \frac{d^2 v_C}{dt^2} + \frac{dv_C}{dt} + \frac{C_2}{C_1}\frac{dv_C}{dt} - \frac{i_1}{C_1}$$

Hence

$$i_1 = R_2 C_2 C_1 \frac{d^2 v_C}{dt^2} + (C_1 + C_2)\frac{dv_C}{dt}$$

Using this to substitute for i_1 in equation [31] gives

$$v = R_1 R_2 C_2 C_1 \frac{d^2 v_C}{dt^2} + R_1 (C_1 + C_2)\frac{dv_C}{dt} + R_2 C_2 \frac{dv_C}{dt}$$

$$+ \frac{(C_1 + C_2)}{C_1} v_C - \frac{1}{C_1}\int i_2 dt$$

But

$$v_C = \frac{1}{C_2}\int i_2 dt$$

Hence, with rearrangement, the equation becomes

$$\frac{d^2 v_C}{dt^2} + \left(\frac{R_1 C_1 + R_1 C_2 + R_2 C_2}{R_1 R_2 C_1 C_2}\right)\frac{dv_C}{dt} + \frac{1}{R_1 R_2 C_1 C_2} v_C$$

$$= \frac{v}{R_1 R_2 C_1 C_2}$$

which is the same as equation [30] derived by nodal analysis.

Example 4

Derive the relationship between the output, the potential difference across the inductor of v_L, and the input v for the circuit shown in Fig. 2.15.

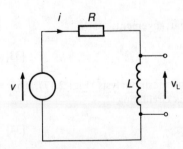

Fig. 2.15 Example 4

Answer

Applying Kirchoff's second law to the circuit loop, then

$$v = v_R + v_L$$

where v_R is the potential difference across the resistor R and v_L that across the inductor. Since according to equations [31] $v_R = iR$, then

$$v = iR + v_L$$

Since, according to equation [18],

$$i = \frac{1}{L}\int v_L \mathrm{d}t$$

then

$$v = \frac{R}{L}\int v_L \mathrm{d}t + v_L$$

Example 5

Derive the relationship between the output, the potential difference across the capacitor C of v_C, for the input v for the circuit shown in Fig. 2.16.

Answer

Using nodal analysis, node B is taken as the reference node and node A taken to be at a potential of v_A relative to B. Applying Kirchoff's law 1 to node A gives

$$i_1 = i_2 + i_3$$

But

$$i_1 = \frac{v - v_A}{R}$$

$$i_2 = \frac{1}{L}\int v_A \mathrm{d}t$$

$$i_3 = C\frac{\mathrm{d}v_A}{\mathrm{d}t}$$

Hence

$$\frac{v - v_A}{R} = \frac{1}{L}\int v_A \mathrm{d}t + C\frac{\mathrm{d}v_A}{\mathrm{d}t}$$

But $v_C = v_A$. Hence, with some rearrangement,

$$v = RC\frac{\mathrm{d}v_C}{\mathrm{d}t} + v_C + \frac{R}{L}\int v_C \mathrm{d}t \qquad [33]$$

The answer can also be obtained by mesh analysis (Fig. 2.17). Thus for the loop in which a current i_1 is circulating

$$v = Ri_1 + L\frac{\mathrm{d}(i_1 - i_2)}{\mathrm{d}t} \qquad [34]$$

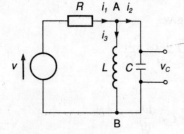

Fig. 2.16 Example 5

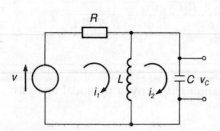

Fig. 2.17 Example 5

For the loop in which the current i_2 is circulating

$$0 = v_C + L\frac{d(i_2 - i_1)}{dt}$$ [35]

Hence

$$v_C = -L\frac{d(i_2 - i_1)}{dt} = L\frac{d(i_1 - i_2)}{dt}$$ [36]

Thus equation [34] becomes

$$v = Ri_1 + v_C$$ [37]

Integrating equation [36] gives

$$\int v_C dt = -L(i_2 - i_1)$$

But

$$i_2 = C\frac{dv_C}{dt}$$

Hence

$$\int v_C dt = -LC\frac{dv_C}{dt} + Li_1$$

and so

$$i_1 = \frac{1}{L}\int v_C dt + C\frac{dv_C}{dt}$$

Thus, substituting for i_1 in equation [37]

$$v = \frac{R}{L}\int v_C dt + RC\frac{dv_C}{dt} + v_C$$

which is the same as equation [30] arrived at by nodal analysis.

Electrical and mechanical analogies The building blocks for electrical and mechanical systems have many similarities. For example, the electrical resistor does not store energy but dissipates it with the current i through the resistor being given by

$$i = \frac{v}{R}$$

where R is a constant, and the power P dissipated by

$$P = \frac{v^2}{R}$$

The mechanical analogue of the resistor is the dashpot. It also does not store energy but dissipates it with the force F being related to the velocity v by

$$F = cv$$

Table 2.3 Analogous electrical and mechanical building blocks

Building block	Describing equation	Energy/ power	AC*
Energy storage			
Inductor	$i = \dfrac{1}{L}\displaystyle\int v\,dt$	$E = \tfrac{1}{2}Li^2$	$\dfrac{1}{L}$
Translational spring	$F = kx = k\displaystyle\int v\,dt$	$E = \dfrac{1}{2}\dfrac{F^2}{k}$	k
Torsional spring	$T = k\theta = k\displaystyle\int \omega\,dt$	$E = \dfrac{1}{2}\dfrac{T^2}{k}$	k
Capacitor	$i = C\dfrac{dv}{dt}$	$E = \tfrac{1}{2}Cv^2$	C
Mass	$F = m\dfrac{d^2x}{dt^2} = m\dfrac{dv}{dt}$	$E = \tfrac{1}{2}mv^2$	m
Moment of inertia	$T = I\dfrac{d^2\theta}{dt^2} = I\dfrac{d\omega}{dt}$	$E = \tfrac{1}{2}I\omega^2$	I
Energy dissipation			
Resistor	$i = \dfrac{v}{\omega s}$	$P = \dfrac{1}{R}v^2$	$\dfrac{1}{R}$
Translational dashpot	$F = cv$	$P = cv^2$	c
Rotational damper	$T = c\omega$	$P = c\omega^2$	c

* Analogous constant.

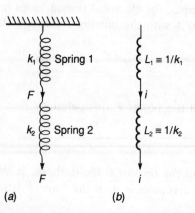

(a)   (b)

Fig. 2.18 Mechanical and electrical analogies

where c is a constant, and the power P dissipated by

$$P = cv^2$$

Both these sets of equations have similar forms. Comparing them, if we take the current as being analogous to the force then the potential difference is analogous to the velocity and the dashpot constant c to the reciprocal of the resistance, i.e. $(1/R)$. These analogies between current and force, potential difference and velocity hold for the other building blocks. Table 2.3 shows the analogous equations.

Consider the electrical analogy for two springs in series, as in Fig. 2.18(a). When a force F is applied to the arrangement, then the force acting on each spring is the same, namely F. The electrical equivalent of force is current i and the equivalents of springs are inductors. Since the same force is experienced by each spring then the same current must be experienced by each inductor. This can only be the case if the two equivalent inductors are in series (Fig. 2.18(b)). For spring 1 the equivalent of k_1 is an inductance of $1/L_1$, for spring 2 k_2 is equivalent to $1/L_2$.

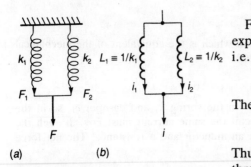

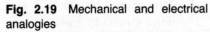

Fig. 2.19 Mechanical and electrical analogies

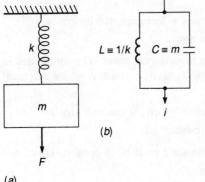

Fig. 2.20 Mechanical and electrical analogies

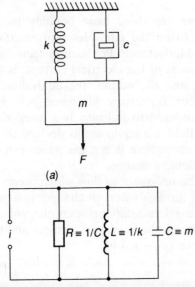

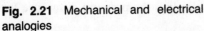

Fig. 2.21 Mechanical and electrical analogies

For two springs in parallel (Fig. 2.19(a)) the forces experienced by each spring must be equal to the total force F, i.e.

$$F = F_1 + F_2$$

The equivalent of this is

$$i = i_1 + i_2$$

Thus the total current must equal the sum of the currents through the equivalent inductors. This can only be the case when the inductors are in parallel (Fig. 2.19(b)). For spring 1 the equivalent of k_1 is an inductance of $1/L_1$, for spring 2 k_2 is equivalent to $1/L_2$.

Figure 2.20(a) shows a mechanical system involving a spring and a mass. The net force acting on the mass is

Net force acting on mass
$$= F - \text{net force exerted by spring}$$

Thus

$$F = \text{net force exerted by spring} + \text{net force acting on mass}$$

This has an electrical equivalent, since an inductor is equivalent to a spring and a capacitor to a mass, of

$$i = \text{current through inductor} + \text{current through capacitor}$$

This can only be the case if the components are in parallel. Thus the electrical analogy is as shown in Fig. 2.20(b).

Figure 2.21 shows a mechanical system involving a spring, a dashpot and a mass. The net force acting on the mass is

Net force acting on mass $= F -$ force exerted by spring
$$- \text{force exerted by dashpot}$$

Thus

$$F = \text{net force acting on mass} + \text{force exerted by spring}$$
$$+ \text{force exerted by dashpot}$$

This has the electrical equivalent, since an inductor is equivalent to a spring, a capacitor to a mass and a resistor to a dashpot, of

$$i = \text{current through capacitor} + \text{current through inductor}$$
$$+ \text{current through dashpot}$$

This can only be the case if the components are in parallel. Thus the electrical analogy is as shown in Fig. 2.21(b).

The analogy between current and force is the one most often used. However, another set of analogies can be drawn between potential difference and force.

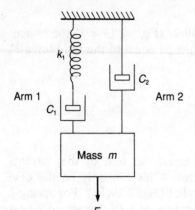

Fig. 2.22 Example 6

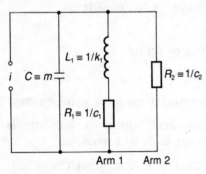

Fig. 2.23 Example 6

Fluid system building blocks

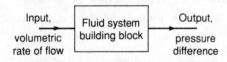

Fig. 2.24 Fluid system building block

Example 6

Draw the electrical circuit which is the equivalent of the mechanical system shown in Fig. 2.22.

Answer

The same force will act on the spring k_1 and dashpot c_1, so in the equivalent electrical circuit the same current must flow through the equivalent components, an inductor and a resistance. The net force acting on the mass is

Net force acting on mass = F − force exerted by arm 1

− force exerted by arm 2

Hence

F = net force acting on mass + force exerted by arm 1

+ force exerted by arm 2

The electrical equivalent of the mass is a capacitor. The component in arm 2 is a dashpot and it will have a resistor as its electrical equivalent. Hence

i = current through capacitor + current through arm 1

+ current through resistance 2

The capacitor, arm 1 and resistance 2 must be in parallel. Hence the circuit is as shown in Fig. 2.23.

In fluid-flow systems there are three basic building blocks which can be considered to be the equivalents of electrical resistance, capacitance and inductance. For such systems (Fig. 2.24) the input, the equivalent of the electrical current, is the volumetric rate of flow q and the output, the equivalent of electrical potential difference, is pressure difference $(p_1 - p_2)$. Fluid systems can be considered to fall into two categories: *hydraulic* ones where the fluid is a liquid and is deemed to be incompressible and *pneumatic* where it is a gas which can be compressed and shows a density change.

Hydraulic resistance is the resistance to flow which occurs as a result of a liquid flowing through valves or changes in a pipe diameter (Fig. 2.25). The relationship between the volume rate of flow of liquid q through the resistance element and the resulting pressure difference $(p_1 - p_2)$ is

$$p_1 - p_2 = Rq \qquad [38]$$

where R is a constant called the hydraulic resistance. The bigger is the resistance the bigger is the pressure difference for a given rate of flow.

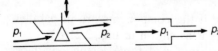

Fig. 2.25 Hydraulic resistance examples

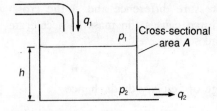

Fig. 2.26 Hydraulic capacitance

Hydraulic capacitance is the term used to describe energy storage with a liquid, where it is stored in the form of potential energy. A height of liquid in a container (Fig. 2.26), i.e. a so-called pressure head, is one form of such a storage. For such a capacitance, the rate of change of volume V in the container, i.e. dV/dt, is equal to the difference between the volumetric rate at which liquid enters the container q_1 and the rate q_2 at which it leaves.

$$q_1 - q_2 = \frac{dV}{dt}$$

But $V = Ah$, where A is the cross-sectional area of the container and h the height of liquid in it. Hence

$$q_1 - q_2 = \frac{d(Ah)}{dt} = A\frac{dh}{dt}$$

But the pressure difference between the input and output is p, where

$$p = h\rho g$$

where ρ is the liquid density and g the acceleration due to gravity. Thus

$$q_1 - q_2 = A\frac{d(p/\rho g)}{dt} = \frac{A}{\rho g}\frac{dp}{dt}$$

if the liquid is assumed to be incompressible, i.e. its density does not change with pressure. The hydraulic capacitance C is defined as being

$$C = \frac{A}{\rho g} \qquad [39]$$

Thus

$$q_1 - q_2 = C\frac{dp}{dt} \qquad [40]$$

Integration of this equation gives

$$p = \frac{1}{C}\int (q_1 - q_2)dt \qquad [41]$$

Hydraulic inertance is the equivalent of inductance in electrical systems or a spring in mechanical systems. To accelerate a fluid and so increase its velocity a force is required. Consider a block of liquid of mass m (Fig. 2.27). The net force acting on the liquid is

$$F_1 - F_2 = p_1A - p_2A = (p_1 - p_2)A$$

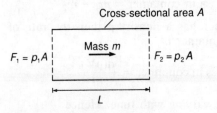

Fig. 2.27 Hydraulic inertance

where $(p_1 - p_2)$ is the pressure difference and A the cross-sectional area. This net force causes the mass to accelerate with an acceleration a, and so

$$(p_1 - p_2)A = ma$$

But a is the rate of change of velocity dv/dt, hence

$$(p_1 - p_2)A = m\frac{dv}{dt}$$

But the mass of liquid concerned has a volume of AL, where L is the length of the block of liquid or the distance between the points in the liquid where the pressures p_1 and p_2 are measured. If the liquid has a density ρ then $m = AL\rho$, and so

$$(p_1 - p_2)A = AL\rho\,\frac{dv}{dt}$$

But the volume rate of flow $q = Av$, hence

$$(p_1 - p_2)A = L\rho\,\frac{dq}{dt}$$

$$p_1 - p_2 = I\frac{dq}{dt} \qquad [42]$$

where the hydraulic inertance I is defined as

$$I = \frac{L\rho}{A} \qquad [43]$$

With pneumatic systems the three basic building blocks are, as with hydraulic systems, resistance, capacitance and inertance. However, gases differ from liquids in being compressible, i.e. a change in pressure causes a change in volume and hence density. *Pneumatic resistance R* is defined in terms of the mass rate of flow $\dot{m}$ and the pressure difference $(p_1 - p_2)$ as

$$p_1 - p_2 = R\dot{m} \qquad [44]$$

Pneumatic capacitance C is due to the compressibility of the gas, rather like the way compression of a spring stores energy. If there is a mass rate of flow $\dot{m}_1$ entering a container of volume V and a mass rate of flow of $\dot{m}_2$ leaving it, then the rate at which the mass in the container is changing is

$$\text{Rate of change of mass in container} = \dot{m}_1 - \dot{m}_2$$

If the gas in the container has a density ρ then the rate of change of mass in the container is

$$\text{Rate of change of mass in container} = \frac{d(\rho V)}{dt}$$

But both ρ and V can be varying with time. Hence

$$\text{Rate of change of mass in container} = \frac{\rho\,\mathrm{d}V}{\mathrm{d}t} + \frac{V\,\mathrm{d}\rho}{\mathrm{d}t}$$

Since $(\mathrm{d}V/\mathrm{d}t) = (\mathrm{d}V/\mathrm{d}p)(\mathrm{d}p/\mathrm{d}t)$ and for an ideal gas $pV = mRT$, with consequently $p = (m/V)RT = \rho RT$ and so $\mathrm{d}\rho/\mathrm{d}t = (1/RT)(\mathrm{d}p/\mathrm{d}t)$, then

$$\text{Rate of change of mass in container} = \rho\,\frac{\mathrm{d}V}{\mathrm{d}p}\frac{\mathrm{d}p}{\mathrm{d}t} + \frac{V}{RT}\frac{\mathrm{d}p}{\mathrm{d}t}$$

where R is the gas constant and T the temperature, assumed constant, on the Kelvin scale. Thus

$$\dot{m}_1 - \dot{m}_2 = \left(\rho\,\frac{\mathrm{d}V}{\mathrm{d}p} + \frac{V}{RT} \right)\frac{\mathrm{d}p}{\mathrm{d}t}$$

The pneumatic capacitance due to the change in volume of the container C_1 is defined as

$$C_1 = \rho\,\frac{\mathrm{d}V}{\mathrm{d}p} \qquad [45]$$

and the pneumatic capacitance due to the compressibility of the gas C_2 as

$$C_2 = \frac{V}{RT} \qquad [46]$$

Hence

$$\dot{m}_1 - \dot{m}_2 = (C_1 + C_2)\frac{\mathrm{d}p}{\mathrm{d}t} \qquad [47]$$

or

$$p_1 - p_2 = \frac{1}{C_1 + C_2}\int (\dot{m}_1 - \dot{m}_2)\mathrm{d}t \qquad [48]$$

Pneumatic inertance is due to the pressure drop necessary to accelerate a block of gas. According to Newton's second law

$$\text{Net force} = \frac{\mathrm{d}(mv)}{\mathrm{d}t}$$

Since the force is provided by the pressure difference $(p_1 - p_2)$, then if A is the cross-sectional area of the block of gas being accelerated

$$(p_1 - p_2)A = \frac{\mathrm{d}(mv)}{\mathrm{d}t}$$

But m, the mass of the gas being accelerated, is ρLA with ρ being the gas density and L the length of the block of gas being accelerated. But the volume rate of flow $q = Av$, where v is the velocity. Thus

$$mv = \rho LA(q/A) = \rho Lq$$

and so

$$(p_1 - p_2)A = L\frac{d(\rho q)}{dt}$$

But $\dot{m} = \rho q$, and so

$$p_1 - p_2 = \frac{L}{A}\frac{d\dot{m}}{dt}$$

$$p_1 - p_2 = I\frac{d\dot{m}}{dt} \qquad [49]$$

with the pneumatic inertance I being

$$I = \frac{L}{A} \qquad [50]$$

Table 2.4 shows the basic characteristics of the fluid building blocks, both hydraulic and pneumatic, and the analogous electrical building blocks. For hydraulics the volumetric rate of

Table 2.4   Analogous fluid and electrical building blocks

Building block	Describing equation	Energy/power	AC*
Energy storage			
Inductor	$i = \frac{1}{L}\int v\,dt$	$E = \frac{1}{2}Li^2$	$\frac{1}{L}$
Hydraulic inertance	$q = \frac{1}{L}\int (p_1 - p_2)dt$	$E = \frac{1}{2}Iq^2$	$\frac{1}{L}$
Pneumatic inertance	$\dot{m} = \frac{1}{L}\int (p_1 - p_2)dt$	$E = \frac{1}{2}I\dot{m}^2$	$\frac{1}{L}$
Capacitor	$i = C\frac{dv}{dt}$	$E = \frac{1}{2}Cv^2$	C
Hydraulic capacitance	$q = C\frac{d(p_1 - p_2)}{dt}$	$E = \frac{1}{2}C(p_1 - p_2)^2$	C
Pneumatic capacitance	$\dot{m} = C\frac{d(p_1 - p_2)}{dt}$	$E = \frac{1}{2}C(p_1 - p_2)^2$	C
Energy dissipation			
Resistor	$i = \frac{v}{R}$	$P = \frac{1}{R}v^2$	$\frac{1}{R}$
Hydraulic resistance	$q = \frac{(p_1 - p_2)}{R}$	$P = \frac{1}{R}(p_1 - p_2)^2$	$\frac{1}{R}$
Pneumatic resistance	$\dot{m} = \frac{(p_1 - p_2)}{R}$	$P = \frac{1}{R}(p_1 - p_2)^2$	$\frac{1}{R}$

* Analogous constant.

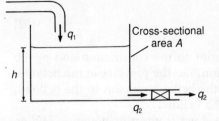

Fig. 2.28 A fluid system

Building up a model for a fluid system

flow and for pneumatics the mass rate of flow are analogous to the electrical current in an electric system. For both hydraulics and pneumatics the pressure difference is analogous to the potential difference in electrical systems. Hydraulic and pneumatic inertance and capacitance are both energy-storage elements, hydraulic and pneumatic resistance are both energy dissipaters.

Figure 2.28 shows a simple hydraulic system: a liquid entering and leaving a container. Such a system can be considered to be a capacitor, the liquid in the container, with a resistor, the valve. Inertance can be neglected since flow rates change only very slowly.

For the capacitor, equation [40] gives

$$q_1 - q_2 = C \frac{dp}{dt}$$

The rate q_2 at which liquid leaves the container equals the rate at which it leaves the valve. Thus for the resistor, equation [38] gives

$$p = Rq_2$$

The pressure is the pressure due to the height of liquid in the container. Thus substituting for q_2 in the first equation gives

$$q_1 - \frac{p}{R} = C \frac{dp}{dt}$$

Since $p = h\rho g$, where ρ is the liquid density and g the acceleration due to gravity, then

$$q_1 - \frac{h\rho r}{R} = C \frac{d(h\rho g)}{dt}$$

and since $C = A/\rho g$, then

$$q_1 = A \frac{dh}{dt} + \frac{\rho g h}{R} \qquad [51]$$

This equation describes how the height of liquid in the container depends on the rate of input of liquid into the container.

A bellows is an example of a simple pneumatic system (Fig. 2.29). Resistance is provided by a constriction which restricts the rate of flow of gas into the bellows and capacitance is provided by the bellows itself. Inertance can be neglected since the flow rate changes only slowly.

The mass flow rate $\dot{m}$ into the bellows is given by equation [44] as

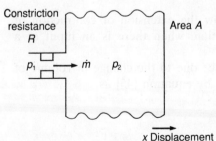

Fig. 2.29 A pneumatic system

$$\dot{m} = \frac{p_1 - p_2}{R} \qquad [52]$$

where p_1 is the pressure prior to the constriction and p_2 the pressure after the constriction, i.e. the pressure in the bellows. All the gas that flows into the bellows remains in the bellows, there being no exit from the bellows.

The capacitance of the bellows is given by equation [49] as

$$\dot{m}_1 - \dot{m}_2 = (C_1 + C_2) \frac{dp_2}{dt}$$

But $\dot{m}_1$ is the mass flow rate $\dot{m}$ given by equation [52] and since there is no exit of gas from the bellows $\dot{m}_2$ is zero. Thus, using the value for the mass flow rate given by equation [52],

$$\frac{p_1 - p_2}{R} = (C_1 + C_2) \frac{dp_2}{dt}$$

Hence

$$p_1 = R(C_1 + C_2) \frac{dp_2}{dt} + p_2 \qquad [53]$$

This equation describes how the pressure in the bellows p_2 varies with time when there is an input of a pressure p_1.

The bellows expands or contracts as a result of pressure changes inside it. Bellows are just a form of spring and so we can write for the relationship between the force F causing an expansion or contraction and the resulting displacement x

$$F = kx$$

where k is the spring constant for the bellows. But the force F depends on the pressure p_2, with $p_2 = F/A$ where A is the cross-sectional area of the bellows. Thus

$$p_2 A = F = kx$$

hence substituting for p_2 in equation [53] gives

$$p_1 = R(C_1 + C_2) \frac{k}{A} \frac{dx}{dt} + \frac{k}{A} x \qquad [54]$$

This equation describes how the extension or contraction of the bellows changes with time when there is an input of a pressure p_1.

The pneumatic capacitance due to the change in volume of the container C_1 is defined by equation [45] as

$$C_1 = \rho \frac{dV}{dp_2}$$

Since $V = Ax$ then

$$C_1 = \rho A \frac{dx}{dp_2}$$

But for the bellows $p_2 A = kx$, thus

$$C_1 = \rho A \frac{dx}{d(kx/A)} = \frac{\rho A^2}{k} \qquad [55]$$

Example 7

Figure 2.30 shows a hydraulic system. Derive relationships which describe how the heights of the liquids in the two containers will change with time. Neglect inertance.

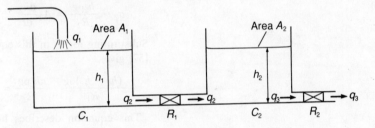

Fig. 2.30 Example 7

Answer

Container 1 is a capacitor and thus, using equation [40]

$$q_1 - q_2 = C_1 \frac{dp}{dt}$$

where $p = h_1 \rho g$ and $C_1 = A_1/\rho g$ and so

$$q_1 - q_2 = A_1 \frac{dh_1}{dt} \qquad [56]$$

The rate at which liquid leaves the container q_2 equals the rate at which it leaves the valve R_1. Thus for the resistor, equation [38] gives

$$p = R_1 q_2$$

The pressure difference p between the two sides of the valve are $h_1 \rho g$ and $h_2 \rho g$. Thus

$$(h_1 - h_2)\rho g = R_1 q_2 \qquad [57]$$

Using the value of q_2 given by this equation and substituting it into equation [56] gives

$$q_1 - \frac{(h_1 - h_2)\rho g}{R_1} = A_1 \frac{dh_1}{dt} \qquad [58]$$

This equation describes how the height of the liquid in container 1 depends on the input rate of flow.

For container 2 a similar set of equations can be derived. Thus for the capacitor C_2

$$q_2 - q_3 = C_2 \frac{dp}{dt}$$

where $p = h_2 \rho g$ and $C_2 = A_2/\rho g$ and so

$$q_2 - q_3 = A_2 \frac{\mathrm{d}h_2}{\mathrm{d}t} \qquad [59]$$

The rate at which liquid leaves the container q_3 equals the rate at which it leaves the valve R_2. Thus for the resistor, equation [38] gives

$$p = R_2 q_3$$

The pressure difference p between the two sides of the valve are $h_2 \rho g$ and 0, assuming the liquid exits into the atmosphere. Thus

$$h_2 \rho g = R_2 q_3 \qquad [60]$$

Using the value of q_3 given by this equation and substituting it into equation [59] gives

$$q_2 - \frac{h_2 \rho g}{R_2} = A_2 \frac{\mathrm{d}h_2}{\mathrm{d}t} \qquad [61]$$

Substituting for q_2 in this equation using the value given by equation [57] gives

$$\frac{(h_1 - h_2)\rho g}{R_1} - \frac{h_2 \rho g}{R_2} = A_2 \frac{\mathrm{d}h_2}{\mathrm{d}t} \qquad [62]$$

This equation describes how the height of liquid in container 2 changes. Thus equations [58] and [62] describe the height variations in the two containers.

Example 8

Figure 2.31 shows a U-tube containing a liquid. Derive an expression indicating how the height difference between the two limbs will vary with time when the pressure above the liquid in one of the limbs is increased and draw the circuit diagram which would be the electrical analogue of the hydraulic system.

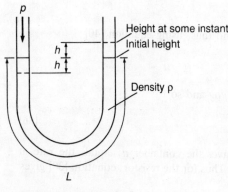

p

h

h

Height at some instant

Initial height

Density ρ

L

Fig. 2.31 Example 8

Answer

When there is no difference in pressure the liquid in both limbs is at the same height. When a pressure p (this being the amount by which the pressure above the liquid in one limb exceeds that above the liquid in the other limb, the so-called gauge pressure when the other limb is open to the atmosphere) occurs then a height difference h between the liquid in the two limbs starts to occur. At any instant of time the pressure difference p between the two limbs must equal the total of the pressure drops across the system. The system can be considered to have inertance, resistance and capacitance and thus p equals the sum of the pressure drops due to each of these. The pressure drop due to inertance is given by equation [42] as

$$\text{Pressure drop} = I \frac{\mathrm{d}q}{\mathrm{d}t}$$

where q is the volume rate of flow of liquid from one limb to the other. The pressure drop due to resistance is given by equation [38] as

$$\text{Pressure drop} = Rq$$

The pressure drop due to the capacitance is given by equation [41] as

Pressure drop $= \dfrac{1}{C} \displaystyle\int q\,\mathrm{d}t$

Thus since p equals the sum of these pressure drops,

$$p = I\frac{\mathrm{d}q}{\mathrm{d}t} + Rq + \frac{1}{C}\int q\,\mathrm{d}t$$

The volume of liquid that has flowed from one limb to the other when there is a height difference of $2h$ is Ah, since a height h of the liquid in one limb can be considered to have been moved from one limb to the other to produce this difference in height. Thus q, the rate at which volume of liquid moves from one limb to the other, is $\mathrm{d}(Ah)/\mathrm{d}t$. Hence

$$p = IA\frac{\mathrm{d}^2h}{\mathrm{d}t^2} + RA\frac{\mathrm{d}h}{\mathrm{d}t} + \frac{A}{C}\int \mathrm{d}h$$

But the total difference in height between the two limbs is $2h$. Thus, since $I = \rho L/A$ and $C = A/\rho g$

$$p = \rho L\frac{\mathrm{d}^2h}{\mathrm{d}t^2} + RA\frac{\mathrm{d}h}{\mathrm{d}t} + 2h\rho g$$

The system has the pressure drops due to the inertance, resistance and capacitance adding up. The electrical equivalent of this is the addition of potential differences across an inductor, resistor and capacitor. This would mean that the three components were in series. The circuit is thus as shown in Fig. 2.32.

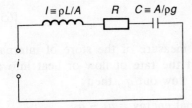

Fig. 2.32 Example 8

Thermal system building blocks

There are only two basic building blocks for thermal systems: resistance and capacitance. There is only a net flow of heat between two points if there is a temperature difference between them. The electrical equivalent of this is that there is only a net current i between two points if there is a potential difference v between them, the relationship between the current and potential difference being

$$i = \frac{v}{R}$$

A similar relationship can be used to define *thermal resistance* R. If q is the rate of flow of heat and $(T_1 - T_2)$ the temperature difference, then

$$q = \frac{(T_2 - T_1)}{R} \qquad [63]$$

The value of the resistance depends on the mode of heat transfer. In the case of conduction through a solid, for unidirectional conduction

$$q = Ak\frac{(T_1 - T_2)}{L}$$

where A is the cross-sectional area of the material through which the heat is being conducted and L the length of material between the points at which the temperatures are T_1 and T_2. k is the thermal conductivity. Hence, comparing this equation with equation [63]

$$R = \frac{L}{Ak} \qquad [64]$$

When the mode of heat transfer is convection, as with liquids and gases, then

$$q = Ah(T_2 - T_1)$$

where A is the surface area across which there is the temperature difference and h is the coefficient of heat transfer. Thus, comparing this equation with equation [63]

$$R = \frac{1}{Ah} \qquad [65]$$

Thermal capacitance is a measure of the store of internal energy in a system. Thus if the rate of flow of heat into a system is q_1 and the rate of flow out q_2, then

Rate of change of internal energy $= q_1 - q_2$

An increase in internal energy means an increase in temperature. Since

Internal energy change $= mc \times$ change in temperature

where m is the mass and c the specific heat capacity, then

Rate of change of internal energy $= mc \times$ rate of change
of temperature

Thus

$$q_1 - q_2 = mc\frac{\mathrm{d}T}{\mathrm{d}t}$$

where $\mathrm{d}T/\mathrm{d}t$ is the rate of change of temperature. This equation can be written as

$$q_1 - q_2 = C\frac{\mathrm{d}T}{\mathrm{d}t} \qquad [66]$$

where C is the thermal capacitance.

$$C = mc \qquad [67]$$

Table 2.5 shows a comparison of the thermal and electrical building blocks. There is no thermal equivalent to the electrical inductor. Electrical resistors dissipate energy, transforming it into heat. Thermal resistance cannot be described as

Table 2.5 Analogous thermal and electrical building blocks

Building block	Describing equation	Energy/power	AC*
Energy storage			
Capacitor	$i = C\dfrac{dv}{dt}$	$E = \tfrac{1}{2}Cv^2$	C
Thermal capacitance	$q_1 - q_2 = C\dfrac{dT}{dt}$	$E = CT$	C
Resistor	$i = \dfrac{v}{R}$	$P = \dfrac{1}{R}v^2$	$\dfrac{1}{R}$
Thermal resistance	$q = \dfrac{(T_1 - T_2)}{R}$	$P = q = \dfrac{(T_1 - T_2)}{R}$	$\dfrac{1}{R}$

* Analogous constant.

an energy dissipater but describes the consequence of there being a temperature difference, just describing a heat flow.

Building up a model for a thermal system

Consider a thermometer at temperature T which has just been inserted into a liquid at temperature T_L (Fig. 2.33). If the thermal resistance to heat flow from the liquid to the thermometer is R then, using equation [63],

$$q = \frac{T_L - T}{R} \tag{68}$$

where q is the net rate of heat flow from liquid to thermometer.

The thermal capacitance C of the thermometer is given by equation [64] as

$$q_1 - q_2 = C\frac{dT}{dt}$$

Since there is only a net flow of heat from the liquid to the thermometer, then $q_1 = q$ and $q_2 = 0$. Thus

$$q = C\frac{dT}{dt} \tag{69}$$

Using this value of q to substitute for it in equation [66],

$$C\frac{dT}{dt} = \frac{T_L - T}{R}$$

Rearranging this equation gives

$$RC\frac{dT}{dt} + T = T_L \tag{70}$$

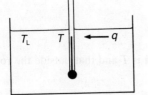

Fig. 2.33 A thermal system

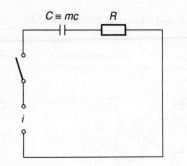

Fig. 2.34 Electrical analogue of Fig. 2.33

This equation describes how the temperature T indicated by the thermometer will vary with time when the thermometer is inserted into a hot liquid. The electrical analogy of this thermal system is the circuit shown in Fig. 2.34, a series capacitor-resistor circuit. Closing the switch is equivalent to the act of inserting the thermometer into the liquid, only then does the current and the heat begin to flow. The change in temperature of the thermometer from its initial value is equivalent to the change in potential difference across the capacitor.

In considering the above thermal system the parameters have been considered to be lumped. This means that, for example, the thermometer has been assumed to be all at the same temperature and the liquid to be all at the same temperature, i.e. the temperatures are only functions of time and not position.

Example 9

Figure 2.35 shows a thermal system consisting of an electric fire in a room. The fire emits heat at the rate q_1 and the room loses heat at the rate q_2. Assuming that the air in the room can be considered to be at a uniform temperature T and that there is no heat storage in the walls of the room, derive an equation describing how the room temperature will change with time.

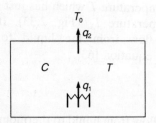

Fig. 2.35 Example 9

Answer

If the air in the room has a thermal capacity C then, using equation [66]

$$q_1 - q_2 = C \frac{dT}{dt}$$

If the temperature inside the room is T and that outside the room T_o then, using equation [63]

$$q_2 = \frac{(T - T_o)}{R}$$

where R is the resistivity of the walls. Substituting for q_2 gives

$$q_1 - \frac{(T - T_o)}{R} = C \frac{dT}{dt}$$

Hence

$$RC \frac{dT}{dt} + T = Rq_1 + T_o$$

Electromechanical elements

Up until now in this chapter the concern has only been with elements which were separately mechanical, electrical, hydraulic, etc. There are, however, some elements that involve more than one type of energy. There are, for instance,

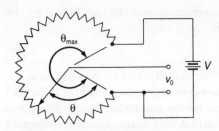

Fig. 2.36 Potentiometer

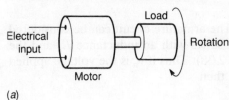

(a)

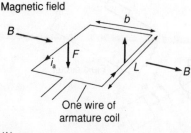

(b)

Fig. 2.37 The d.c. motor: (a) driving a load, (b) basic motor principle

electromechanical devices such as potentiometers, motors and generators. A potentiometer has an input of a rotation and an output of a potential difference, the rotation being used to move the sliding contact over the potentiometer track (Fig. 2.36). An electric motor has an input of a potential difference and an output of rotation of a shaft. A generator has an input of rotation of a shaft and an output of a potential difference.

For the potentiometer (Fig. 2.36), which is a potential divider,

$$\frac{v_{\mathrm{o}}}{V} = \frac{\theta}{\theta_{\max}}$$

where V is the potential difference across the full length of the potentiometer track and $\theta_{\max}$ is the total angle swept out by the slider in being rotated from one end of the track to the other. Thus the ratio of the output v_{o} to the input θ is

$$\frac{\text{output}}{\text{input}} = \frac{v_{\mathrm{o}}}{\theta} = \frac{V}{\theta_{\max}} = \text{a constant} \qquad [71]$$

The d.c. motor is used to convert an electrical input signal into a mechanical output signal (Fig. 2.37(a)). The motor basically consists of a coil, the armature coil, which is free to rotate. This coil is located in the magnetic field provided by, usually, a current through field coils. When a current i_{a} flows through the armature coil then, because it is in a magnetic field, forces act on the coil and cause it to rotate (Fig. 2.37(b)). The force F acting on a wire carrying a current i_{a} and of length L in a magnetic field of flux density B at right angles to the wire is given by

$$F = Bi_{\mathrm{a}}L$$

With N wires

$$F = NBi_{\mathrm{a}}L$$

The forces on the armature coil wires result in a torque T, where $T = Fb$, with b being the breadth of the coil. Thus

$$T = NBi_{\mathrm{a}}Lb$$

The resulting torque is thus proportional to Bi_{a}, the other factors all being constants. Hence we can write

$$T = k_1 Bi_{\mathrm{a}} \qquad [72]$$

Since the armature is a coil rotating in a magnetic field, a voltage will be induced in it as a result of electromagnetic induction. The direction of this voltage will be in such a direction as to oppose the change producing it and is thus called the back e.m.f. This back e.m.f. v_{b} is proportional to the rate or rotation of the armature and the flux linked by the coil, hence the flux density B. Thus

$$v_b = k_2 B \omega \tag{73}$$

where ω is the shaft angular velocity and k_2 is a constant.

With a so-called *armature-controlled motor* the field current i_f is held constant and the motor controlled by adjusting the armature voltage v_a. A constant field current means a constant magnetic flux density B for the armature coil. Thus equation [73] gives

$$v_b = k_3 \omega$$

where k_3 is a constant. The armature circuit can be considered to be a resistance R_a in series with an inductance L_a and the back e.m.f. source (Fig. 2.38). Thus if v_a is the voltage applied to the armature circuit, then

$$v_a - v_b = L_a \frac{di_a}{dt} + R_a i_a \tag{74}$$

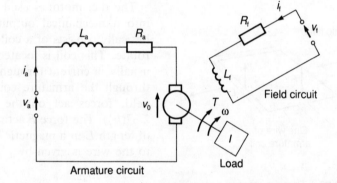

Fig. 2.38 d.c. motor circuits

We can think of this equation in terms of the block diagram shown in Fig. 2.39(a). The input to the motor part of the system is v_a and this is summed with the feedback signal of the back e.m.f. v_b to give an error signal which is the input to the armature circuit. The above equation thus describes the relationship between the input of the error signal to the armature coil and the output of the armature current i_a. Substituting for v_b in equation [74] gives

$$v_a - k_3 \omega = L_a \frac{di_a}{dt} + R_a i_a \tag{75}$$

The current i_a in the armature leads to a torque T, given by equation [72]. Since B is constant, this equation becomes

$$T = k_1 B i_a = k_4 i_a$$

where k_4 is a constant. This torque then becomes the input to

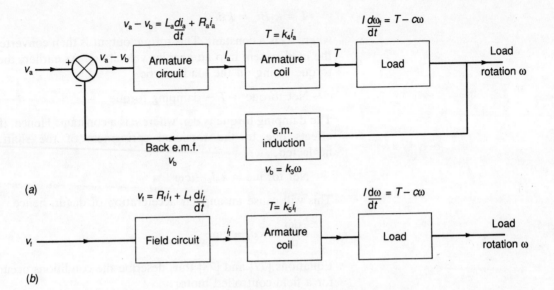

$$v_a - v_b = L_a\frac{di_a}{dt} + R_a i_a$$

$$T = k_4 i_a$$

$$I\frac{d\omega}{dt} = T - c\omega$$

v_a + $v_a - v_b$ **Armature circuit** i_a **Armature coil** T **Load**   Load rotation ω

e.m. induction

Back e.m.f. v_b

$$v_b = k_3\omega$$

(a)

$$v_f = R_f i_f + L_f\frac{di_f}{dt}$$

$$T = k_5 i_f$$

$$I\frac{d\omega}{dt} = T - c\omega$$

v_f **Field circuit** i_f **Armature coil**   **Load**   Load rotation ω

(b)

Fig. 2.39 (a) Armature-controlled, and (b) field-controlled d.c. motors

the load system. The net torque acting on the load will be

Net torque = T − damping torque

The damping torque is $c\omega$, where c is a constant. Hence, if any effects due to the torsional springiness of the shaft are neglected,

Net torque = $k_4 i_a - c\omega$

This will cause an angular acceleration of $d\omega/dt$, hence

$$I\frac{d\omega}{dt} = k_4 i_a - c\omega \qquad [76]$$

Equations [75] and [76] thus describe the conditions occurring for an armature-controlled motor.

With a so-called *field-controlled motor* the armature current is held constant and the motor is controlled by varying the field voltage. For the field circuit (Fig. 2.38) there is essentially just inductance L_f in series with a resistance R_f. Thus, for that circuit,

$$v_f = R_f i_f + L_f\frac{di_f}{dt} \qquad [77]$$

We can think of the field-controlled motor in terms of the block diagram shown in Fig. 2.39(*b*). The input to the system is v_f. The field circuit converts this into a current i_f, the relationship between v_f and i_f being the above equation. This current leads to the production of a magnetic field and hence a torque acting on the armature coil, as indicated by equation [72]. But the flux density B is proportional to the field current i_f and i_a is constant, hence equation [77] can be written as

$$T = k_1 B i_a = k_5 i_f$$

where k_5 is a constant. This torque output is then converted by the load system into an angular velocity ω. As earlier, the net torque acting on the load will be

$$\text{Net torque} = T - \text{damping torque}$$

The damping torque is $c\omega$, where c is a constant. Hence, if any effects due to the torsional springiness of the shaft are neglected,

$$\text{Net torque} = k_5 i_f - c\omega$$

This will cause an angular acceleration of $d\omega/dt$, hence

$$I \frac{d\omega}{dt} = k_5 i_f - c\omega \qquad [78]$$

Equations [77] and [78] thus describe the conditions occurring for a field-controlled motor.

Linearity

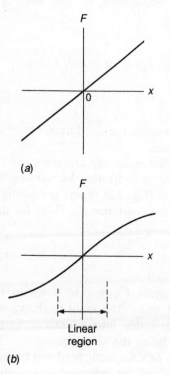

(a)

(b)

Linear region

Fig. 2.40 (a) Ideal spring, (b) real spring

The relationship between the force F and the extension x produced for an ideal spring is linear, being given by $F = kx$. This means that if force F_1 produces an extension x_1 and force F_2 produces extension x_2 then a force equal to $(F_1 + F_2)$ will produce an extension $(x_1 + x_2)$. This is called the *principle of superposition* and is a necessary condition for a linear system. Another condition for a linear system is that if an input F_1 produces an extension x_1 then an input cF_1 will produce an output cx_1, where c is a constant multiplier. A graph of the force F plotted against the extension x is a straight line passing through the origin when the relationship is linear (Fig. 2.40(a)).

Real springs, like any other real components, are however not perfectly linear (Fig. 2.40(b)). However, there is often a range of operation for which linearity can be assumed. Thus for the spring giving the graph in Fig. 2.40(b), linearity can be assumed provided the spring is only used over the central part of its graph. For many system components, linearity can be assumed for operations within a range of values of the variable about some operating point.

For some system components (Fig. 2.41) the relationship is non-linear. For such components the best that can be done to obtain a linear relationship is to consider the slope of the graph at the operating point. Thus for the relationship between y and x in Fig. 2.41, at the operating point P where the slope has the value m

$$\Delta y = m \Delta x \qquad [79]$$

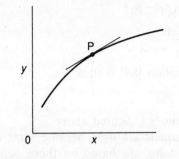

Fig. 2.41 A non-linear relationship

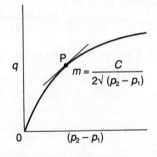

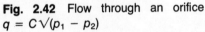

Fig. 2.42 Flow through an orifice $q = C\sqrt{(p_1 - p_2)}$

where Δy and Δx are small changes in input and output signals at the operating point.

Thus, for example, the rate of flow of liquid q through an orifice is given by

$$q = c_d A \sqrt{[2(p_1 - p_2)/\rho]} \qquad [80]$$

where c_d is a constant called the discharge coefficient, A the cross-sectional area of the orifice, ρ the fluid density and $(p_1 - p_2)$ the pressure difference. For a constant cross-sectional area and density the equation can be written as

$$q = C\sqrt{(p_1 - p_2)} \qquad [81]$$

where C is a constant. This is a non-linear relationship between the rate of flow and the pressure difference. We can obtain a linear relationship by considering the slope of the rate of flow/pressure difference graph (Fig. 2.42) at the operating point. The slope is $dq/d(p_1 - p_2)$ and has the value

$$m = \frac{dq}{d(p_1 - p_2)} = \frac{d[C\sqrt{(p_1 - p_2)}]}{d(p_1 - p_2)} = \frac{C}{2\sqrt{(p_{o1} - p_{o2})}} \qquad [82]$$

where $(p_{o1} - p_{o2})$ is the value at the operating point. Thus for small changes about the operating point

$$\Delta q = m\Delta(p_1 - p_2) \qquad [83]$$

where m has the value given by equation [82].

Hence if, for equation [81] we had $C = 2m^3/s$ per kPa, i.e.

$$q = 2\sqrt{(p_1 - p_2)}$$

then for an operating point of $(p_1 - p_2) = 4\,kPa$ the linearized version of the equation, i.e. equation [83], would be, using the value of m given by equation [82],

$$\Delta q = \frac{2}{2\sqrt{4}}\Delta(p_1 - p_2)$$

$$\Delta q = 0.5\,\Delta(p_1 - p_2)$$

In the above discussion it was assumed that the flow was through an orifice of constant cross-sectional area. If the orifice is a flow-control valve then this is not the case, the cross-sectional area being adjusted to vary the flow rate. In such a situation, instead of equation [81] we would have, for equation [80],

$$q = CA\sqrt{(p_1 - p_2)} \qquad [84]$$

and for changes about the operating point the slopes of a graph of q against A would be

$$m_1 = \frac{dq}{dt} = C\sqrt{(p_1 - p_2)}$$

and for a graph of q against $(p_1 - p_2)$

$$m_2 = \frac{dq}{d(p_1 - p_2)} = \frac{CA}{2\sqrt{(p_1 - p_2)}}$$

The linearized version of equation [84] is thus

$$\Delta q = m_1 \Delta A + m_2 \Delta(p_1 - p_2) \qquad [85]$$

with m_1 and m_2 having the values indicated above.

Linearized mathematical models are used because most of the techniques of control systems are based on there being linear relationships for the elements of such systems. Also, because most control systems are maintaining an output equal to some reference value, the variations from this value tend to be rather small and so the linearized model is perfectly appropriate.

Example 10

A thermistor is used for temperature measurement in a control system. The relationship between the resistance R of the thermistor and its temperature T is given by

$$R = k \exp -cT$$

Linearize this equation about an operating point of T_o.

Answer

The slope m of a graph of R against T at the operating point T_o is given by dR/dT.

$$m = \frac{dR}{dT} -kc \exp -cT_o$$

Hence

$$\Delta R = m\Delta T = (-kc \exp -cT_o)T$$

Hydraulic-mechanical elements

Figure 2.43 shows a hydraulic system in which an input displacement x_i is, after passing through a hydraulic system, transformed into a displacement x_o of a load. Such a system is used in the power-assisted steering system of a car and enables a small amount of power used in producing a displacement of the steering wheel to become a larger amount of power which can produce a displacement which turns the wheels of the car.

The system consists of a *spool valve* and a *cylinder*. The input displacement x_i to the left results in the hydraulic fluid supply pressure p_s causing fluid to flow into the left-hand side of the cylinder. This pushes the piston in the cylinder to the right and expels the fluid in the right-hand side of the chamber through the exit port at the right-hand end of the spool valve. The rate of flow of fluid to and from the chamber depends on

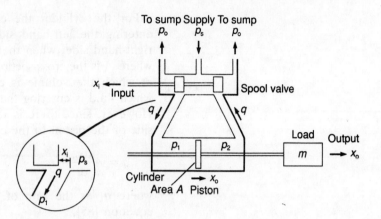

Fig. 2.43 Hydraulic system and load

the extent to which the input motion has uncovered the ports allowing the fluid to enter or leave the spool valve. When the input displacement x_i is to the right the spool valve allows fluid to move to the right-hand end of the cylinder and so results in a movement of the piston in the cylinder to the left.

The rate of flow of fluid q through an orifice, which is what the ports in the spool valve are, is a non-linear relationship (equation [80]) depending on the pressure difference between the two sides of the orifice and the cross-sectional area A of it. A linearized version of the equation is however used, equation [85], i.e.

$$\Delta q = m_1 \Delta A + m_2 \Delta (\text{pressure difference})$$

where m_1 and m_2 are constants at the operating point. For the fluid entering the chamber the pressure difference is $(p_s - p_1)$ and for the exit $(p_2 - p_o)$. If the operating point about which the equation is linearized is taken to be when the spool valve is central and the ports connecting it to the cylinder are both closed then for this condition: q is zero and so $\Delta q = q$, A is proportional to x_s if x_s is measured from this central position, and the change in pressure on the inlet side of the piston is $-\Delta p_1$ relative to p_s and on the exit side Δp_2 relative to p_o. Thus for the inlet port the equation can be written as

$$q = m_1 x_i + m_2(-\Delta p_1)$$

and for the exit port

$$q = m_1 x_i + m_2 \Delta p_2$$

Adding the two equations gives

$$2q = 2m_1 x_i - m_2(\Delta p_1 - \Delta p_2)$$

$$q = m_1 x_i - m_3(\Delta p_1 - \Delta p_2) \qquad [86]$$

where $m_3 = m_2/2$.

For the cylinder the change in the volume of the fluid entering the left-hand side of the chamber, or leaving the right-hand side, when the piston moves a distance x_o is Ax_o, where A is the cross-sectional area of the piston. Thus the rate at which the volume is changing is $A(dx_o/dt)$. The rate at which fluid is entering the left-hand side of the cylinder is q. However, since there is some leakage flow of fluid from one side of the piston to the other

$$q = A\frac{dx_o}{dt} + q_L$$

where q_L is the rate of leakage. Substituting for q, using equation [85],

$$m_1 x_i - m_3(\Delta p_1 - \Delta p_2) = A\frac{dx_o}{dt} + q_L \qquad [87]$$

The rate of leakage flow q_L is a flow through an orifice, the gap between the piston and the cylinder. This is of constant cross-section and a pressure difference $(\Delta p_1 - \Delta p_2)$. Hence using the linearized equation for such a flow, i.e. equation [83],

$$q_L = m_4(\Delta p_1 - \Delta p_2)$$

Thus using this equation to substitute for q_L in equation [87]

$$m_1 x_i - m_3(\Delta p_1 - \Delta p_2) = A\frac{dx_o}{dt} + m_4(\Delta p_1 - \Delta p_2)$$

$$m_1 x_i - (m_3 + m_4)(\Delta p_1 - \Delta p_2) = A\frac{dx_o}{dt} \qquad [88]$$

The pressure difference across the piston results in a force being exerted on the load, the force exerted being $(\Delta p_1 - \Delta p_2)A$. There is however some damping of motion, i.e. friction, of the mass. This is proportional to the velocity of the mass, i.e. (dx_o/dt). Hence the net force acting on the load is

$$\text{Net force} = (\Delta p_1 - \Delta p_2)A - c\frac{dx}{dt}$$

This net force causes the mass to accelerate, the acceleration being (dv/dt) or (d^2x_o/dt^2). Hence

$$m\frac{d^2x_0}{dt^2} = (\Delta p_1 - \Delta p_2)A - c\frac{dx_o}{dt}$$

Rearranging this equation gives

$$\Delta p_1 - \Delta p_2 = \frac{m}{A}\frac{d^2x_o}{dt^2} + \frac{c}{A}\frac{dx_o}{dt}$$

Substituting for the pressure difference in equation [88],

$$m_1 x_i - (m_3 + m_4)\left(\frac{m}{A}\frac{d^2 x_o}{dt^2} + \frac{c}{A}\frac{dx_o}{dt}\right) = A\frac{dx_o}{dt}$$

Rearranging this gives

$$\frac{(m_3 + m_4)m}{A}\frac{d^2 x_o}{dt^2} + \left(A + \frac{c(m_3 + m_4)}{A}\right)\frac{dx_o}{dt} = m_1 x_1 \qquad [89]$$

This equation can be simplified by introducing two constants k and τ, this latter constant being called the time constant (see Chapter 3). Rearranging equation [89] leads to

$$\frac{(m_3 + m_4)m}{A}\frac{d^2 x_o}{dt^2}$$

$$+ \left(\frac{A^2 + c(m_3 + m_4)}{A}\right)\frac{dx_o}{dt} = m_1 x_i$$

$$\frac{(m_3 + m_4)m}{A^2 + c(m_3 + m_4)}\frac{d^2 x_o}{dt^2} + \frac{dx_o}{dt} = \frac{Am_1}{A^2 + c(m_3 + m_4)}x_i$$

$$\tau\frac{d^2 x_o}{dt^2} + \frac{dx_o}{dt} = k x_i \qquad [90]$$

where

$$\tau = \frac{(m_3 + m_4)m}{A^2 + c(m_3 + m_4)}$$

and

$$k = \frac{Am_i}{A^2 + c(m_3 + m_4)}$$

Problems

1 Derive an equation relating the input, force F, with the output, displacement x, for the systems described by Fig. 2.44.

2 Propose a model for a stepped shaft, i.e. a shaft where there is a change in diameter, used to rotate a mass and derive an equation relating the input torque and the angular rotation. You may neglect damping.

3 Derive the relationship between the output, the potential difference v_R across the resistor R, and the input v for the circuit shown in Fig. 2.45 which has a resistor in series with a capacitor.

4 Derive the relationship between the output, the potential difference v_R across the resistor R, and the input v for the series LCR circuit shown in Fig. 2.46.

5 Derive the relationship between the output, the potential difference v_C across the capacitor C, and the input v for the circuit shown in Fig. 2.47.

6 Draw the electrical analogous circuits for the mechanical systems shown in Fig. 2.48.

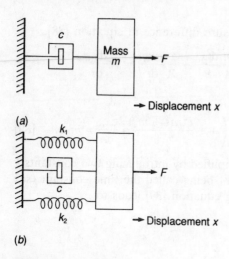

(a)

(b)

Fig. 2.44 Problem 1

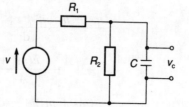

Fig. 2.45 Problem 3

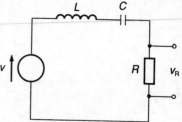

Fig. 2.46 Problem 4

Fig. 2.47 Problem 5

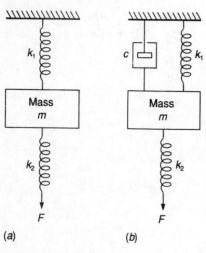

(a)   (b)

Fig. 2.48 Problem 6

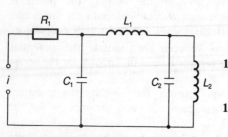

Fig. 2.49 Problem 7

7 Draw the mechanical analogue for the electrical circuit shown in Fig. 2.49.

8 Derive the relationship between the height h_2 and time for the hydraulic system shown in Fig. 2.50. Neglect inertance.

9 Figure 2.51 shows two linked, identical open containers. Derive a relationship indicating how the height of liquid in a container will vary if there is 'sloshing' of liquid between the two containers as a result of a disturbance in one container, such as an increase in pressure above the liquid surface.

10 A hot object, capacitance C, and temperature T cools in a large room at temperature T_r. If the thermal system has a resistance R derive an equation describing how the temperature of the hot object changes with time and give an electrical analogue of the system.

11 Figure 2.52 shows a thermal system involving two compartments, with one containing a heater. If the temperature of the compartment containing the heater is T_1, the temperature of the other compartment T_2 and the temperature surrounding the compartments T_3, develop equations describing how the temperatures T_1 and T_2 will vary with time. All the walls of the containers have the same resistance and negligible capacity. The two containers have the same capacity C.

12 The relationship between the force F used to stretch a spring and its extension x is given by $F = kx^2$, where k is a constant. Linearize this equation for an operating point of x_0.

13 The relationship between the e.m.f. E generated by a thermocouple and the temperature T is given by

$$E = aT + bT^2$$

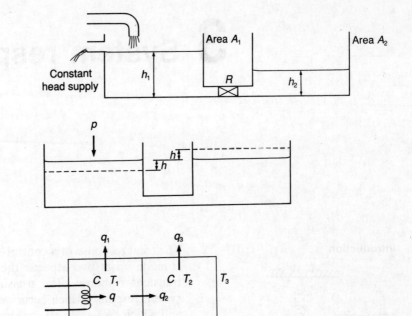

Fig. 2.50 Problem 8

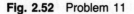

Fig. 2.51 Problem 9

Fig. 2.52 Problem 11

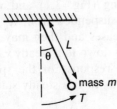

Fig. 2.53 Problem 14

where a and b are constants. Linearize this equation for an operating point of temperature T_o.

14 The relationship between the torque T applied to a simple pendulum and the angular deflection θ (Fig. 2.53) is given by

$$T = MgL \sin \theta$$

where m is the mass of the pendulum bob, L the length of the pendulum and g the acceleration due to gravity. Linearize this equation for the equilibrium angle θ of $0°$.

3 System response

Introduction

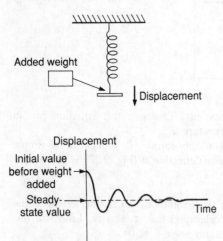

Fig. 3.1 Transient and steady-state responses of a spring system

The total response of a control system, or element of a system, is made up of two aspects: the steady-state response and the transient response. The *transient response* is that part of a system response which occurs when there is a change in input and which dies away after a short interval of time. The *steady-state response* is the response that remains after all transient responses have died down. To give a simple illustration of this, consider a vertically suspended spring (Fig. 3.1) and what happens when a weight is suddenly suspended from it. The deflection of the spring abruptly increases and then may well oscillate until after some time it settles down to a steady value. The steady value is the steady-state response of the spring system; the oscillation that occurs prior to this steady state is the transient response.

The input to the spring system is a quantity which varies with time; this input is the weight. Up to some particular time there is no added weight, then after that time there is. This type of input is known as a step input. The output from the system is a quantity which varies with time. Both the input and the output are functions of time. One way of indicating this is to write them in the form $f(t)$, where f is the function and (t) indicates that its value depends on time t. Note that $f(t)$ does not mean that f should be multiplied by t. Thus for the weight W input to the spring system we could write $W(t)$ and for the deflection d output $d(t)$. A block diagram of the system would thus look like Fig. 3.2.

To describe the behaviour of a system fully a model must describe the relationship between inputs and outputs which are functions of time and thus be able to describe both the transient and steady-state behaviours. A model is thus needed which will indicate how the system response will vary with time. One type of model that is often used to describe the behaviour of a control system or control system element is a *differential equation*. Such a model includes derivatives with

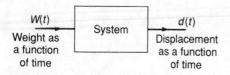

Fig. 3.2 The spring system

respect to time and so gives information about how the response of a system varies with time. A derivative dx/dt describes the rate at which x varies with time, the derivative d^2x/dt^2 (or $d(dx/dt)/dt$) states how dx/dt varies with time.

Differential equations are equations involving derivatives. They can be classed as *first order*, *second order*, *third order*, etc., according to the highest order of the derivative in the equation. For a first-order equation the highest order will be dx/dt, with a second-order d^2x/dt^2, with a third-order d^3x/dt^3, with an nth order d^nx/dt^n.

This chapter is about the types of responses we can expect from first-order and second-order systems and the solution of such differential equations in order that the response of the system to different types of input can be obtained. Essentially, the methods used can be classified as either trying a solution to find one that fits or transforming the equation into another form that can be handled by conventional algebra. This chapter looks at the 'try a solution' approach, with Chapter 4 dealing with the transformation method.

Examples of first-order systems

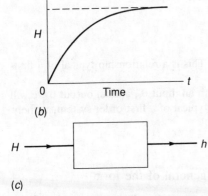

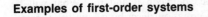

(a)

(b)

(c)

Fig. 3.3 (a) A float-controlled water tank, (b) the water height variation with time, (c) the system with its input and output

An example of a first-order system is a float-controlled water tank (Fig. 3.3(a)). For such a system the rate at which water enters the tank, and hence the rate at which the height of the water in the tank changes with time, depends on the difference in the height h of the water in the tank from the height H at which the float completely switches off the water, i.e.

Rate of change of height is proportional to $(H - h)$

Hence

$$\frac{dh}{dt} = k(H - h)$$

where dh/dt is the rate of change of height and k a constant.

The more the water level rises in the tank the smaller the value of $(H - h)$ and so the smaller becomes the rate of change of height with time (dh/dt). A graph of water height against time will look like that in Fig. 3.3(b). The equation describing this graph is

$$h = H(1 - e^{-kt})$$

Such a system can be considered to have as an input the required height H and an output h (Fig. 3.3(c)).

Another example of a first-order system is a capacitor in series with a resistor (Fig. 3.4). The rate of change of the potential difference v_C across the capacitor with time, i.e. dv_C/dt, is proportional to the difference in value between v_C and the input voltage to the system V.

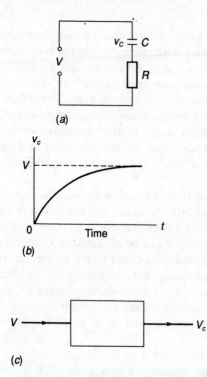

(a)

(b)

(c)

Fig. 3.4 (a) Series RC circuit, (b) the variation of the potential difference across the capacitor with time, (c) the system and its input and output

Rate of change of v_C is proportional to $(V - v_C)$

This is usually written as

$$\frac{dv_C}{dt} = \frac{1}{RC}(V - v_C)$$

where R is the resistance and C the capacitance. Figure 3.4(b) shows how v_C varies with time. The graph has the equation

$$v_C = V(1 - e^{-t/RC})$$

Such a system can be considered to have as an input the voltage V and as an output the potential difference v_C (Fig. 3.4(c)).

All first-order systems have the characteristic that the rate of change of some variable is proportional to the difference between this variable and some set value of the variable (this could, however, be zero).

Example 1

When a thermometer is placed in a hot liquid at some temperature θ_H the rate at which the reading θ of the thermometer changes with time is proportional to the difference between θ and θ_H. What will be (a) the form of the differential equation describing how the thermometer temperature changes with time and (b) the equation of a graph of θ with time?

Answer

(a) The system has

Rate of change of θ proportional to $(\theta_H - \theta)$

The differential equation is thus

$\frac{d\theta}{dt}$ is proportional to $(\theta_H - \theta)$

$$\frac{d\theta}{dt} = k(\theta_H - \theta)$$

where k is some constant. This is a relationship typical of a first-order system.

(b) For a first-order system with an input θ_H and an output θ we will have the type of equation typical of a first-order system, namely

$$\theta = \theta_H(1 - e^{-kt})$$

The first-order differential equation

A first-order equation is in general of the form

Rate of change of output signal is proportional to $(b_0\theta_i - a_0\theta_o)$

or, as it is more usually written,

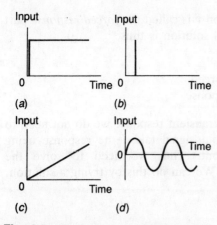

Fig. 3.5 Input signals: (a) step, (b) impulse, (c) ramp, and (d) sinusoidal

$$a_1 \frac{d\theta_o}{dt} + a_0\theta_o = b_0\theta_i \qquad [1]$$

where a_1, a_0 and b_0 are constants, θ_i is the input function to the system and θ_o the output. $d\theta_o/dt$ is the rate at which the output is changing with time.

The input signals to systems can take many forms. A common form is a step input. This is when the input abruptly changes in value, as in Fig. 3.5(a). An example of this is when the voltage is suddenly switched on in a circuit, e.g. the series CR circuit in Fig. 3.4(a). Other forms of input that are often encountered are impulse, ramp and sinusoidal signals (Fig. 3.5). An impulse is a very short duration input, a ramp is a steadily increasing input and a sinusoidal input is one whose oscillations can be described by an equation of the form $\sin \omega t$, with ω being the angular frequency.

Solving a first-order differential equation

There are a number of methods for solving a first-order differential equation; one general method is as follows. For equation [1]

$$a_1 \frac{d\theta_o}{dt} + a_0\theta_o = b_0\theta_i$$

if we make the substitution

$$\theta_o = u + v$$

then

$$a_1 \frac{d(u + v)}{dt} + a_0(u + v) = b\theta_i$$

Rearranging this gives

$$\left(a_1 \frac{du}{dt} + a_0u\right) + \left(a_1 \frac{dv}{dt} + a_0v\right) = b\theta_i$$

If we let

$$a_1 \frac{dv}{dt} + a_0v = b\theta_i \qquad [2]$$

then we must have

$$a_1 \frac{du}{dt} + a_0u = 0 \qquad [3]$$

This last equation, when solved, will give a value for u in an equation for which there is no input. For this reason u is called the *transient* or *free response* part of the solution. The equation involving v has the input function and so gives a value for v when there is the input which forces a particular response from the system. For this reason v is called the *forced response* part of the solution. The total solution is thus

$$\theta_o = \quad u \quad + \quad v$$

Transient   Forced
response   response

In order to obtain the transient response we do not need to know what form any input signal takes, the response being independent of the input. Thus we need to solve the differential equation [3]. We can do this by trying a solution. Try

$$u = A\,e^{st}$$

Then, since $du/dt = As\,e^{st}$, the differential equation becomes

$$a_1 As\,e^{st} + a_0 A\,e^{st} = 0$$

and so, since the $A\,e^{st}$ terms cancel,

$$a_1 s + a_0 = 0$$

$$s = -\frac{a_0}{a_1}$$

Thus

$$u = A\exp -(a_0 t/a_1)$$

To obtain the solution of the differential equation [2]

$$a_1\frac{dv}{dt} + a_0 v = b\theta_i$$

we also try a solution. The form of the solution to try will depend on the form of the input signal. Thus for a step input when θ_i is a constant for times greater than $t = 0$ the solution to try is $v = k$, where k is a constant. Where we have a signal that can be described by an equation of the form $\theta_i = a + bt + ct^2 + \ldots$, where any of the coefficients a, b, c, etc., may be zero, then we try a solution of the form $v = a + bt + ct^2 + \ldots$. Hence for a ramp signal where we have $\theta_i = bt$ then the solution to try is $v = bt$. For a sinusoidal, or cosine, signal we try $v = A\sin\omega t + B\cos\omega t$.

Suppose we take the input to be a step input (Fig. 3.5(a)) occurring at $t = 0$, i.e. the input suddenly at $t = 0$ jumps to a value θ_i and remains constant at that value for the rest of time then we can try as a possible solution

$$v = k$$

Since k is a constant $dv/dt = 0$ and so the differential equation becomes

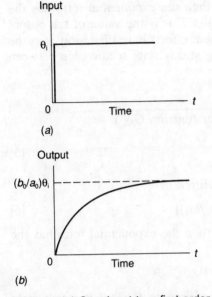

(a)

(b)

Fig. 3.6 (a) Step input to a first-order system and (b) the resulting output

$$a_0 k = b_0 \theta_i$$

and so

$$v = \frac{b_0 \theta_i}{a_0}$$

The full solution is thus

$$\theta_o = u + v$$

$$\theta_o = A \exp -(a_0 t / a_1) + (b_0 / a_0)\theta_i$$

We can determine the value of the constant A given some initial (boundary) condition. Thus if $\theta_o = 0$ when $t = 0$, then

$$0 = A + (b_0 / a_0)\theta_i$$

Thus $A = -(b_0 / a_0)\theta_i$ and so the equation becomes

$$\theta_o = -(b_0 / a_0)\theta_i \exp -(a_0 t / a_1) + (b_0 / a_0)\theta_i$$

$$\theta_o = (b_0 / a_0)\theta_i [1 - \exp(-a_0 t / a_1)] \qquad [4]$$

Figure 3.6(b) shows a graph of how the output θ_o varies with time for the step input shown in Fig. 3.6(a). The graph and the equation are general and describe the response of all first-order systems to a step input which occurs at $t = 0$.

Example 2

An electrical system, a resistance in series with a capacitor (as in Fig. 3.4), when subject to a step input of size V is found to give an output of a potential difference across the capacitor v_C which is given by the differential equation

$$RC \frac{dv_C}{dt} + v_C = V$$

What is the solution of the differential equation, i.e. how does v_C vary with time?

Answer

Comparing the differential equation with equation [1]

$$a_1 \frac{d\theta_o}{dt} + a_0 \theta_o = b_0 \theta_i$$

then $a_1 = RC$, $a_0 = 1$, and $b_0 = 1$. Then the solution is of the form given by equation [4]

$$\theta_o = (b_0 / a_0)\theta_i [1 - \exp -(a_0 t / a_1)]$$

$$v_C = V[1 - \exp -(t / RC)]$$

The time constant

For the solution, equation [4], of a first-order differential equation [1]

$$\theta_o = (b_0/a_0)\theta_i[1 - \exp-(a_0t/a_1)] \qquad [4]$$

When $t = 0$ then the exponential term has the value of 1 and thus $\theta_o = 0$. When $t = \infty$ then the exponential term has the value 0 and so $a_o = (b_0/a_0)\theta_i$. This is the value of the output that occurs when the transient effects have died away, i.e. the steady-state value. Thus the steady state relationship between the output and the input is

$$\theta_o = (b_0/a_0)\theta_i$$

and the steady-state transfer function G_{SS} is

$$G_{SS} = \frac{\theta_0}{\theta_i} = \frac{b_0}{a_0} \qquad [5]$$

Equation [4] can thus be written as

$$\theta_o = G_{SS}\theta_i[1 - \exp-(a_0t/a_1)] \qquad [6]$$

When the time $t = (a_1/a_0)$ then the exponential term has the value $e^{-t} = 0.37$ and

$$\theta_o = G_{SS}\theta_i(1 - 0.37) = 0.63G_{SS}\theta_i$$

In this time the output has risen to 0.63 of it steady-state value. This time is called the *time constant* τ. In a time of $2(a_1/a_0) = 2\tau$, then the exponential term becomes $e^{-2} = 0.14$ and so

$$\theta_o = G_{SS}\theta_i(1 - 0.14) = 0.86G_{SS}\theta_i$$

In a similar way, values can be calculate for the output after 3τ, 4τ, 5τ, etc. Table 3.1 shows the results of such calculations and Fig. 3.7 the graph of θ_o/θ_i plotted against time. These results are general and apply to all systems giving first-order differential equations.

Thus equation [6] can be written in terms of the time constant as

$$\theta_o = G_{SS}\theta_o(1 - e^{-t/\tau}) \qquad [7]$$

and we can use the relationships $G_{SS} = b_0/a_0$ and $\tau = a_1/a_0$ to write the first-order differential equation (equation [1]) in the form

$$a_1\frac{d\theta_0}{dt} + a_0\theta_o = b_0\theta_i$$

$$\frac{a_1d\theta_o}{a_0dt} + \theta_o = \frac{b_0}{a_0}\theta_i$$

$$\tau\frac{d\theta_o}{dt} + \theta_o = G_{SS}\theta_i \qquad [8]$$

Table 3 1 Response of a first-order system to a step input

Time t	θ_o/θ_i
0	0
1τ	$0.63G_{SS}$
2τ	$0.86G_{SS}$
3τ	$0.95G_{SS}$
4τ	$0.98G_{SS}$
5τ	$0.99G_{SS}$
∞	G_{SS}

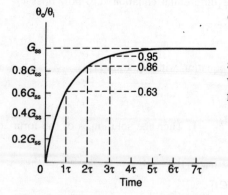

Fig. 3.7 Response of a first-order system to a step input

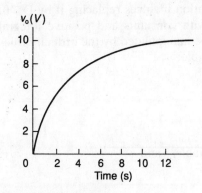

$v_o(V)$

Fig. 3.8 Example 3

Example 3

Figure 3.8 shows how the output v_o of a first-order system varies with time when subject to a step input of 5 V. Estimate (a) the time constant, (b) the steady-state transfer function and (c) the first-order differential equation for the system.

Answer

(a) The time constant τ is the time taken for a first-order system output to change from 0 to 0.63 of its final steady-state value. In this case this time is about 3 s. We can check this value, and that the system is first order, by finding the value at 2τ, i.e. 6 s. With a first-order system it should be 0.86 of the steady-state value. In this case it is.

(b) The steady-state output is 10 V. Thus the steady-state transfer function G_{SS} is (steady-state output/input) = 10/5 = 2.

(c) The differential equation for a first-order system can be written as (equation [8])

$$\tau \frac{d\theta_o}{dt} + \theta_o = G_{SS}\theta_i$$

$$3\frac{dv_o}{dt} + v_o = 2v_i$$

Example 4

The relationship between the output angular velocity ω_o and the input voltage v_i for a motor when subject to a step input is found to be given by

$$\frac{JR}{K_1K_2} \frac{d\omega_o}{dt} + \omega_o = \frac{1}{K_1}v_i$$

What will be the steady-state value of the angular velocity and the time constant of the system?

Answer

The differential equation can be compared with equation [8]

$$\tau \frac{d\theta_o}{dt} + \theta_o = G_{SS}\theta_i$$

Thus the steady-state transfer function G_{SS} is $1/K_1$ and hence the steady-state angular velocity $(1/K_1)v_i$. The time constant is JR/K_1K_2.

D operator

One way of considering differential equations is in terms of what is called the *D operator*. In a differential equation $d\theta/dt$ is replaced by $D\theta$, with D being called the *operator*. Similarly $d^2\theta/dt^2$ can be replaced by $D^2\theta$. In general

$$D^n\theta = \frac{d^n\theta}{dt^n} \qquad [9]$$

Integration of θ in an equation involves replacing it by $D^{-1}\theta$. The D, in combinations with constants and positive integral powers of itself, can then be manipulated by the ordinary rules of algebra. Thus, for example:

$$D\theta + \theta = (D + 1)\theta$$

where θ is the variable;

$$D(u + v) = Du + Dv$$

where u and v are the variables; and

$$D^nD^m\theta = D^{n+m}\theta$$

where θ is the variable and n and m are positive integers.

D operator and a first-order equation

A first-order differential equation (equation [1])

$$a_1\frac{d\theta_o}{dt} + a_0\theta_o = b_0\theta_i$$

becomes, when written in terms of the D operator,

$$a_1D\theta_o + a_0\theta_o = b_0\theta_i \tag{10}$$

There are a number of procedures that can be used to solve differential equations in this form. One method is to follow a similar technique to that described earlier in this chapter for the solution of the differential equation in which the transient and forced responses were found.

Thus let

$$\theta_o = u + v$$

then

$$D\theta_o = Du + Dv$$

and thus equation [10] becomes

$$a_1(Du + Dv) + a_0(u + v) = b_0\theta_i$$

$$(a_1Du + a_0u) + (a_1Dv + a_0v) = b_0\theta_i$$

If

$$a_1Dv + a_0v = b_0\theta_i \tag{11}$$

then

$$a_1Du + a_0u = 0 \tag{12}$$

u is the solution when there is no input, i.e. the transient solution. Suppose we try a solution to this transient equation of the form

$$u = A\,e^{st}$$

then

$$\frac{du}{dt} = Du = As\,e^{st}$$

Hence equation [12] becomes

$$(a_1s + a_0)A\,e^{st} = 0$$

and so

$$(a_1s + a_0) = 0$$

This equation is called the *auxiliary equation*. It is effectively formed by taking the differential equation for θ_o in D operator form, equating θ_i to zero and then replacing the differential operator by an algebraic variable s. Hence this means that $s = -a_0/a_1$ and the transient solution is

$$u = A\,e^{st} = A\exp(-a_0t/a_1)$$

To obtain the solution of the differential equation [11]

$$a_1Dv + a_0v = b\theta_i$$

we also try a solution. The form of the solution to try will depend on the form of the input signal. Thus for a step input when θ_i is a constant for times greater than $t = 0$ the solution to try is $v = k$, where k is a constant. Where we have a signal that can be described by an equation of the form $\theta_i = a + bt + ct^2 + \ldots$, where any of the coefficients a, b, c, etc., may be zero, then we try a solution of the form $v = a + bt + ct^2 + \ldots$. Hence for a ramp signal where we have $\theta_i = bt$ then the solution to try is $v = bt$. For a sinusoidal, or cosine, signal we try $v = A\sin\omega t + B\cos\omega t$.

Suppose we take the input to be a step input occurring at $t = 0$, i.e. the input suddenly at $t = 0$ jumps to a value θ_i and remains constant at that value for the rest of time then we can try as a possible solution

$$v = k$$

Since k is a constant $dv/dt = Dv = 0$ and so the differential equation becomes

$$a_0k = b_0\theta_i$$

and so

$$v = \frac{b_0}{a_0}\theta_i$$

The full solution is thus

$$\theta_o = u + v$$

$$\theta_o = A\exp-(a_0t/a_1) + (b_0/a_0)\theta_i$$

We can determine the value of the constant A given some initial (boundary) condition. Thus if $\theta_o = 0$ when $t = 0$, then

$$0 = A + (b_0/a_0)\theta_i$$

Thus $A = -(b_0/a_0)\theta_i$ and so the equation becomes

$$\theta_0 = -(b_0/a_0)\theta_i \exp{-(a_0t/a_1)} + (b_0/a_0)\theta_i$$

$$\theta_o = (b_0/a_0)\theta_i[1 - \exp{(-a_0t/a_1)}] \qquad [13]$$

The steady-state value of the output θ_o is obtained when there are no changes of the output with time, i.e. $d\theta_o/dt = D\theta_o = 0$. When this occurs then $\theta_o/\theta_i = b_0/a_0$. Thus we can define a steady-state transfer function G_{SS} as

$$G_{SS} = \frac{\theta_o}{\theta_1} = \frac{b_0}{a_0}$$

a_1/a_0 is called the time constant τ (see earlier this chapter) and so equation [13] can be written as

$$\theta_o = G_{SS}\theta_i(1 - e^{-t/\tau}) \qquad [14]$$

The D operator equation [10] for the first-order system

$$a_1D\theta_o + a_0\theta_o = b_0\theta_i$$

is often written in the form

$$\frac{\theta_o}{\theta_i} = \frac{b_0}{a_1D + a_0}$$

$$\frac{\theta_o}{\theta_i} = \frac{b_0/a_0}{(a_1/a_0)D + 1}$$

and with the time constant τ as a_1/a_0 and the steady-state transfer function G_{SS} as b_0/a_0

$$\frac{\theta_o}{\theta_i} = \frac{G_{SS}}{\tau D + 1} \qquad [15]$$

This form of equation is typical of a first-order system.

Example 5

The displacement d of a bellows (Fig. 3.9) is related to the gauge pressure p, i.e. the pressure in the bellows relative to the atmospheric pressure, by a first-order differential equation

$$\frac{d}{p} = \frac{k}{\tau D + 1}$$

Solve this equation for a step input of pressure at $t = 0$.

Answer

This equation is of the form given in equation [15]

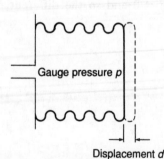

Gauge pressure p

Displacement d

Fig. 3.9 Example 5

$$\frac{\theta_o}{\theta_i} = \frac{G_{SS}}{\tau D + 1}$$

for which the solution, when there was a step input at $t = 0$, is equation [14]

$$\theta_o = G_{SS}\theta_i(1 - e^{-t/\tau})$$

Thus the solution for the bellows is

$$d = kp(1 - e^{-t/\tau})$$

Examples of second-order systems

Many second-order systems can be considered to be essentially just a stretched spring with a mass and some means of providing damping. Figure 3.10 shows the basis of such a system. The forces acting on the mass m are the applied force F in one direction and, in the opposite direction, the force exerted by the stretched spring and by the damping. The force due to the spring is proportional to the amount x by which the spring has been stretched and can thus be written as kx where k is a constant for the spring. The force due to the damping, effectively a piston moving in a container, is proportional to the rate at which the displacement of the piston is changing, i.e. proportional to dx/dt. Thus the force due to the damping can be written as $c\,dx/dt$, where c is a constant. Hence the net force acting on the mass is

$$\text{Net force} = F - kx - c\frac{dx}{dt}$$

This net force causes an acceleration a of the mass (Newton's second law). Thus

$$F - kx - c\frac{dx}{dt} = ma$$

But acceleration is the rate of change of velocity dv/dt and velocity v is the rate of change of displacement x, i.e. dx/dt. Thus

$$a = \frac{dv}{dt} = \frac{d(dx/dt)}{dt} = \frac{d^2x}{dt^2}$$

Hence

$$F - kx - c\frac{dx}{dt} = m\frac{d^2x}{dt^2}$$

or, when rearranged,

$$m\frac{d^2x}{dt^2} + c\frac{dx}{dt} + kx = F$$

This is a second-order differential equation of a system which has had an abrupt application of the force F, i.e. a step

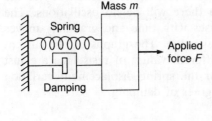

Mass m

Spring

Damping

Applied force F

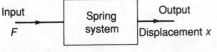

Input F — Spring system — Output Displacement x

Fig. 3.10 A second-order spring system

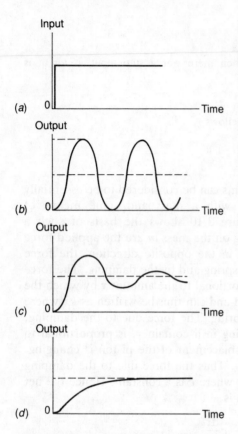

Fig. 3.11 Types of output variation with time (a) for a step input with (b) no damping, (c) some damping, (d) high damping.

input. The way in which the resulting displacement x will vary with time will depend on the amount of damping in the system. If there was no damping at all then the mass would freely oscillate on the spring and the oscillations would continue indefinitely. No damping means $c\,dx/dt = 0$. However damping will cause the oscillations to die away until a steady displacement of the mass is obtained. If the damping is high enough there will be no oscillations and the displacement of the mass will just slowly increase with time and gradually the mass will move towards its steady displacement position (Fig. 3.11).

Another example of a second-order system is a series RLC circuit (Fig. 3.12). For such a circuit, when subject to a step input of size V at $t = 0$, the current i is given by

$$\frac{d^2 i}{dt^2} + \frac{R}{L}\frac{di}{dt} + \frac{1}{LC}i = \frac{V}{LC}$$

The damping in the RLC circuit is provided by the resistance. In the absence of the resistance the current in the circuit would oscillate freely and continue oscillating indefinitely. Then the term $(R/L)di/dt$ is zero. However, the presence of resistance results in the oscillations dying away with time and a steady current being obtained. However, if the resistance, i.e. the damping, is large enough there will be no oscillations. The current just slowly increases with time and gradually moves towards the steady current value. The graphs of the current variation with time for different values of resistance are just like those in Fig. 3.11 for the spring displacement variation with time for different degrees of damping.

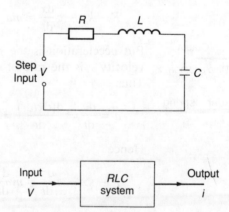

Fig. 3.12 Series RLC circuit

The second-order differential equation

A second-order equation has the general form

$$a_2 \frac{d^2\theta_o}{dt^2} + a_1 \frac{d\theta_o}{dt} + a_0\theta_o = b_0\theta_i \qquad [16]$$

where b_o, a_o, a_1 and a_2 are constants.

In the absence of any damping or forcing the output of a second-order system is a continuous oscillation which we can describe by the equation

$$\theta_o = A \sin \omega_n t$$

where A is the amplitude of the oscillation and ω_n the angular frequency of the free undamped oscillations. Differentiating this gives

$$\frac{d\theta_o}{dt} = \omega_n A \cos \omega_n t$$

$$\frac{d^2\theta_o}{dt^2} = -\omega_n^2 A \sin \omega n t = -\omega_n^2 \theta_0$$

and so the differential equation becomes

$$\frac{d^2\theta_o}{dt^2} + \omega_n^2 \theta_o = 0$$

The general second-order differential equation is thus usually written as

$$\frac{d^2\theta_o}{dt^2} + 2\zeta\omega_n \frac{d\theta_o}{dt} + \omega_n^2\theta_o = b_1\omega_n^2\theta_i \qquad [17]$$

where ω_n is the angular frequency with which the system will freely oscillate in the absence of any damping and ζ is the *damping ratio*. When ζ is 0 then we have the freely oscillating situation described in Fig. 3.11(*b*); when ζ is less than 1 we have that in Fig. 3.11(*c*); and when ζ is greater than 1 we have that in Fig. 3.11(*c*). The reasons for this will become apparent in the next section of this chapter.

Solving a second-order differential equation

A second-order differential equation can be solved by the same method as described earlier for the first-order differential equation. Thus if we substitute for θ_o

$$\theta_o = \quad u \quad + \quad v$$

$$\text{Transient} \quad \text{Forced}$$
$$\text{response} \quad \text{response}$$

then for u

$$\frac{d^2u}{dt^2} + 2\zeta\omega_n \frac{du}{dt} + \omega_n^2 u = 0 \qquad [18]$$

and for v

$$\frac{d^2v}{dt^2} + 2\zeta\omega_n \frac{dv}{dt} + \omega_n^2 v = \omega_n^2 b_0\theta_i \qquad [19]$$

To solve the transient equation we can try a solution of the form

$$u = A e^{st}$$

With such a solution

$$\frac{du}{dt} = As e^{st}$$

$$\frac{d^2u}{dt^2} = As^2 e^{st}$$

Thus equation [18] becomes

$$As^2 e^{st} + 2\zeta\omega_n As e^{st} + \omega_n^2 A e^{st} = 0$$

$$s^2 + 2\zeta\omega_n s + \omega_n^2 = 0 \qquad [20]$$

Thus $u = A e^{st}$ can only be a solution provided the above equation equals 0. This equation is called the *auxiliary equation*. The roots of the equation can be obtained by factorizing or using the formula for the roots of a quadratic equation

$$s = \frac{-b \pm \sqrt{(b^2 - 4ac)}}{2a}$$

with a, b and c being the constants in an equation of the form $(ax^2 + bx + c) = 0$. Thus in the context of the constants given in equation [20]

$$s = \frac{-2\zeta\omega_n \pm \sqrt{(4\zeta^2\omega_n^2 - 4\omega_n^2)}}{2}$$

$$s = -\zeta\omega_n \pm \omega_n \sqrt{(\zeta^2 - 1)} \qquad [21]$$

The values of s obtained from the above equation depends very much on the value of the square root term. Thus when ζ^2 is greater than 1 the square root term gives a square root of a positive number, and when ζ^2 is less than 1 we have the square root of a negative number. The damping factor in determining whether the square root term is that of a positive or negative number is a crucial factor in determining the form of the output from the system.

When $\zeta > 1$ then there are two different real roots s_1 and s_2, where

$$s_1 = -\zeta\omega_n + \omega_n\sqrt{(\zeta^2 - 1)}$$

$$s_2 = -\zeta\omega_n - \omega_n\sqrt{(\zeta^2 - 1)}$$

and so the general solution for u is

$$u = A \exp s_1 t + B \exp s_2 t \qquad [22]$$

For such conditions the system is said to be *overdamped*.

When $\zeta = 1$ there are two equal roots with $s_1 = s_2 = -\omega_n$. For this condition, which is called *critically damped*,

$$u = (At + B)\exp{-\omega_n t} \qquad [23]$$

It may seem that the solution for this case should be $u = A\,e^{st}$, but such a solution, with just one constant A, is not capable of satisfying the initial conditions for a second-order system.

When $\zeta < 1$ there are two complex roots since the roots both involve $\sqrt{(-1)}$. Thus equation [21] can be written as

$$s = -\zeta\omega_n \pm \omega_n\sqrt{(\zeta^2 - 1)}$$

$$s = -\zeta\omega_n \pm \omega_n\sqrt{(-1)(1 - \zeta^2)}$$

and so, writing j for $\sqrt{(-1)}$,

$$s = -\zeta\omega_n \pm j\omega_n\sqrt{(1 - \zeta^2)}$$

If we let

$$\omega = \omega_n\sqrt{(1 - \zeta^2)} \qquad [24]$$

then

$$s = -\zeta\omega_n \pm j\omega$$

and so the two roots are

$$s_1 = -\zeta\omega_n + j\omega$$

$$s_2 = -\zeta\omega_n - j\omega$$

The term ω is the angular frequency of the motion when it is in the damped condition specified by ζ. The solution under these conditions is

$$u = A\exp(-\zeta\omega_n + j\omega)t + B\exp(-\zeta\omega_n + j\omega)t$$

$$u = \exp(-\zeta\omega_n t)[A\exp(j\omega t) + B\exp(-j\omega t)]$$

But

$$e^{j\omega t} = \cos\omega t + j\sin\omega t$$

and

$$e^{-j\omega t} = \cos\omega t - j\sin\omega t$$

Hence

$$u = \exp(-\zeta\omega_n t)(A\cos\omega t + jA\sin\omega t + B\cos\omega t - jB\sin\omega t)$$

$$u = \exp(-\zeta\omega_n t)[(A + B)\cos\omega t + j(A - B)\sin\omega t]$$

If we substitute constants P and Q for $(A + B)$ and $j(A - B)$, then

$$u = \exp(-\zeta\omega_n t)(P\cos\omega t + Q\sin\omega t) \qquad [25]$$

For such conditions the system is said to be *underdamped*.

To solve the forcing equation [18], we consider a particular form of input signal and then try a solution. Thus for a step input of size θ_i at time $t = 0$ then we can try a solution

$$v = k$$

where k is a constant (see the discussion of the solution of first-order differential equations for a discussion of the choice of solutions). For such a solution

$$\frac{dv}{dt} = 0$$

and

$$\frac{d^2v}{dt^2} = 0$$

and so equation [19] becomes

$$0 + 0 + \omega_n^2 k = b_0 \omega_n^2 \theta_i$$

Thus

$$v = b_0 \theta_i$$

The complete solution is thus the sum of u and v, i.e. the transient plus forcing responses. Thus for the overdamped system

$$\theta_o = A \exp s_1 t + B \exp s_2 t + b_0 \theta_i \qquad [26]$$

for the critically damped system

$$\theta_o = (At + B) \exp -\omega_n t + b_0 \theta_i \qquad [27]$$

and for the underdamped system

$$\theta_o = \exp(-\zeta\omega_n t)(P \cos \omega t + Q \sin \omega t) + b_0 \theta_i \qquad [28]$$

When $t \rightarrow \infty$ then the above three equations all lead to the solution $\theta_o = b_0 \theta_i$. Thus

$$\text{Steady-state transfer function } G_{SS} = \frac{\theta_o}{\theta_i} = b_0 \qquad [29]$$

Figure 3.13 shows graphs of the output as a function of time for different degrees of damping, i.e. different values of the damping factor ζ. The time axis is $\omega_n t$. This has been used so that the graphs fit second order systems regardless of their value of ω_n. As a consequence of this the $\theta_o = 0$ values for the undamped oscillation occur for $\omega_n t = 0$, 2π, 4π, etc. with $\theta_o = 2$ values at π, 3π, etc.

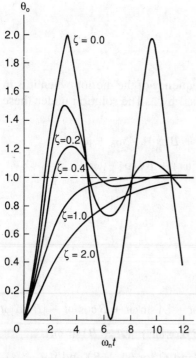

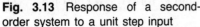

Fig. 3.13 Response of a second-order system to a unit step input

Example 6

A series *RLC* circuit, as in Fig. 3.12, has $R = 100\,\Omega$, $L = 2.0\,H$ and C

$= 20\,\mu F$. The current i in the circuit is given by

$$\frac{d^2i}{dt^2} + \frac{R}{L}\frac{di}{dt} + \frac{1}{LC}i = \frac{V}{LC}$$

when there is a step input V.

(a) What is the natural frequency of the circuit?
(b) Is the system over damped, critically damped or under damped?
(c) What is the damped oscillation frequency?
(d) What is the solution to the differential equation, if when $t = 0$ we have $\theta_o = 0$ and $d\theta_o/dt = 0$?

Answer

(a) In comparing the equation with equation [17] then

$$\frac{d^2\theta_o}{dt^2} + 2\zeta\omega_n\frac{d\theta_o}{dt} + \omega_n^2\theta_o = b_1\omega_n^2\theta_o$$

$$\omega_n^2 = \frac{1}{LC} = \frac{1}{2.0 \times 20 \times 10^{-6}}$$

$$\omega_n = 158\,Hz$$

(b) Comparison with equation [17] gives

$$2\zeta\omega_n = \frac{R}{L} = \frac{100}{2.0}$$

Thus

$$\zeta = \frac{50}{2 \times 158} = 0.16$$

Since ζ is less than 1 the system is underdamped.

(c) The damped oscillation frequency ω is given by equation [24] as

$$\omega = \omega_n(1 - \zeta^2) = 158\sqrt{(1 - 0.18)} = 143\,Hz$$

(d) Because the system is underdamped the solution will be of the form given by equation [28]

$$\theta_o = \exp(-\zeta\omega_n t)(P\cos\omega t + Q\sin\omega t) + b_0\theta_i$$

Since $\theta_o = 0$ when $t = 0$, then

$$0 = 1(P + 0) + b_0\theta_i$$

Thus $P = -b_0\theta_i$.
  Since $d\theta_o/dt = 0$ when $t = 0$ then, since $d(uv)/dt = udv/dt + vdu/dt$,

$$\frac{d\theta_o}{dt} = \exp(-\zeta\omega_n t)(\omega P\sin\omega t - \omega Q\cos\omega t)$$

$$-\zeta\omega_n\exp(-\zeta\omega_n t)(P\cos\omega t + Q\cos\omega t)$$

$$0 = 1(0 - \omega Q) - \zeta\omega_n(P + 0)$$

$$Q = \zeta(\omega_n/\omega)(-b_0\theta_i)$$

Thus

$$\theta_o = \exp(-\zeta\omega_n t)[-b\theta_i\cos\omega t - \zeta(\omega_n/\omega)\theta_i] + b_0\theta_i$$

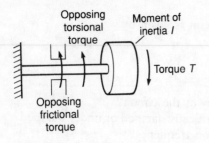

Fig. 3.14 Example 7

Since $\omega_n = 158\,\text{Hz}$, $\zeta = 0.16$, $\omega = 143\,\text{Hz}$, $b_0 = 1$ and $\theta_i = V$, then

$$\theta_o = -V\exp(-25t)[\cos 143t + 0.18\sin 143t] + V$$

Example 7

There are many systems which involve the twisting of a shaft as a result of some input. Figure 3.14 shows the basic system. The input, a torque T, is applied to a disk with a moment of inertia I about the axis of the shaft. The shaft is free to rotate at the disk end but fixed at its far end. The rotation is opposed by the torsional stiffness of the shaft, an opposing torque of $k\theta_o$ occurring for an input rotation of θ_o. k is a constant. Frictional forces damp the rotation of the shaft and provide an opposing torque of $c\,d\theta_o/dt$, where c is a constant and $d\theta_o/dt$ the rate at which the input rotation changes with time.

(a) What is the net torque acting on the shaft when there is an input rotation of θ_o?

(b) A net torque causes an angular acceleration of $I\,d^2\theta_o/dt^2$. Hence write the differential equation for this system.

(c) What is the condition for this system to be critically damped?

Answer

(a) The net torque is

$$\text{Net torque} = T - c\frac{d\theta_o}{dt} - k\theta_o$$

(b) The net torque is $I\,d^2\theta_o/dt$, hence

$$I\frac{d^2\theta_o}{dt^2} = T - c\frac{d\theta_o}{dt} - k\theta_o$$

$$I\frac{d^2\theta_o}{dt^2} + c\frac{d\theta_o}{dt} + k\theta_o = T$$

(c) The condition for critical damping is given by the damping ratio $\zeta = 1$. Comparing the above differential equation with equation [17]

$$\frac{d^2\theta_o}{dt^2} + 2\zeta\omega_n\frac{d\theta_o}{dt} + \omega_n^2\theta_o = b_1\omega_n^2\theta_i$$

Then

$$\omega_n^2 = \frac{k}{I}$$

and

$$2\zeta\omega_n = \frac{c}{I}$$

Thus

$$\zeta = \frac{c}{2\omega_n I} = \frac{c}{2I\sqrt{(k/I)}} = \frac{c}{2\sqrt{(kI)}}$$

Thus critical damping means that we must have

$$c = 2\sqrt{(kI)}$$

Example 8

The output θ_o of a second-order system varies with time when subject to a step input θ_i in accordance with the differential equation

$$\frac{d^2\theta_o}{dt} + 4\frac{d\theta_o}{dt} + 4 = 4\theta_i$$

What is (a) the undamped angular frequency, (b) the damping ratio and (c) the solution of the differential equation when there is a step input of θ_i at $t = 0$, given that at $t = 0$ we have $\theta_o = 0$ and $d\theta_o/dt = 0$?

Answer

(a) In comparing the equation with equation [17]

$$\frac{d^2\theta_o}{dt^2} + 2\zeta\omega_n\frac{d\theta_o}{dt} + \omega_n^2\theta_o = b_1\omega_n^2\theta_i$$

$$\omega_n^2 = 4$$

$$\omega_n = 2\,\text{Hz}$$

(b) In comparing the equation with equation [17]

$$2\zeta\omega_n = 4$$

$$\zeta = 1.0$$

The system is critically damped.

(c) The solution to a critically damped system subject to a step input is of the form given by equation [27], i.e.

$$\theta_o = (At + B)\exp -\omega_n t + b_0\theta_i$$

Since $\theta_o = 0$ when $t = 0$ then

$$0 = (0 + B)1 + b_0\theta_i$$

$$B = -b_0\theta_i$$

Since $d\theta_o/dt = 0$ when $t = 0$, then since $d(uv)/dt = u\,dv/dt + v\,du/dt$,

$$\frac{d\theta_o}{dt} = A\exp -\omega_n t + (At + B)(-\omega_n\exp -\omega_n t)$$

$$0 = A - B\omega_n$$

Hence $A = B\omega_n = -\omega_n b_0\theta_i$. Thus

$$\theta_o = (-\omega_n b_0\theta_i t - b_0\theta_i)\exp -\omega_n t + b_0\theta_i$$

$$\theta_o = -b_0\theta_i(\omega_n t + 1)\exp -\omega_n t + b_0\theta_i$$

Since $\omega_n = 2$ and $b_0 = 1$, then

$$\theta_o = -\theta_i(2t + 1)e^{-2t} + \theta_i$$

Example 9

A second-order system has an output θ_0 which varies with time t when subject to a step input θ_i according to the differential equation

$$\frac{d^2\theta_o}{dt} + 13\frac{d\theta_o}{dt} + 36\theta_o = 36\theta_i$$

(a) What is the undamped angular frequency?
(b) What is the damping ratio?
(c) What is the solution to the differential equation if when $t = 0$ we have $\theta_o = 0$ and $d\theta_o/dt = 0$?

Answer

(a) In comparing the equation with equation [17]

$$\frac{d^2\theta_o}{dt^2} + 2\zeta\omega_n\frac{d\theta_o}{dt} + \omega_n^2\theta_o = b_1\omega_n^2\theta_1 \qquad [17]$$

$$\omega_n^2 = 36$$

$$\omega_n = 6\,\text{Hz}$$

(b) In comparing the equation with equation [17]

$$2\zeta\omega_n = 13$$

$$\zeta = 1.1$$

The system is overdamped.

(c) The equation of an overdamped system when subject to a step input is given by equation [26] as

$$\theta_o = A\exp s_1 t + B\exp s_2 t + b_0\theta_i$$

s_1 and s_2 are the roots of the auxiliary equation obtained by effectively replacing $d^2\theta_o/dt^2$ by s^2, $d\theta_o/dt$ by s and equating the input to 0. Thus

$$s^2 + 13s + 36 = 0$$

$$(s + 3)(s + 12) = 0$$

Thus $s = -3$ or -12. Hence

$$\theta_o = A\,e^{-3t} + B\,e^{-12t} + b_0\theta_i$$

Since $\theta_o = 0$ when $t = 0$, then

$$0 = A + B + b_0\theta_i$$

Since $d\theta_o/dt = 0$ when $t = 0$, then

$$\frac{d\theta_o}{dt} = -3A\,e^{-3t} - 12B\,e^{-12t}$$

$$0 = -3A - 12B$$

Thus $A = -4B$ and so $A = -(4/3)b_0\theta_i$ and $B = (1/3)b_0\theta_i$. Hence, since $b_0 = 1$

$$\theta_o = \theta_i[(1/3)\,e^{-3t} - (4/3)\,e^{-12t} + 1]$$

Performance measures for second-order systems

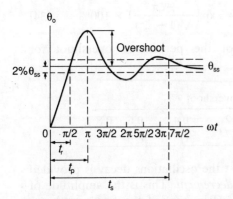

Fig. 3.15 Step response of an under-damped system

Figure 3.15 shows the typical form of the response of an underdamped system to a step input. Certain terms are used to specify such a performance.

The *rise time* t_r is the time taken for the response θ_o to rise from 0 to the steady-state value θ_{SS} and is a measure of how fast a system responds to the input. This is the time for the oscillating response to complete a quarter of a cycle, i.e. $\frac{1}{2}\pi$. Thus

$$\omega t_r = \tfrac{1}{2}\pi \qquad [30]$$

The rise time is sometimes specified as the time taken for the response to rise from some specified percentage of the steady-state value, e.g. 10%, to another specified percentage, e.g. 90%.

The *peak time* t_p is the time taken for the response to rise from 0 to the first peak value. This is the time for the oscillating response to complete a half-cycle, i.e. π. Thus

$$\omega t_p = \pi \qquad [31]$$

The *overshoot* is the maximum amount by which the response overshoots the steady-state value. It is thus the amplitude of the first peak. The overshoot is sometimes written as a percentage of the steady-state value.

Equation [28] gives the response variation with time,

$$\theta_o = \exp(-\zeta\omega_n t)(P\cos\omega t + Q\sin\omega t) + b_0\theta_i$$

Since $\theta_o = 0$ when $t = 0$, then

$$0 = 1(P + 0) + b_0\theta_i$$

$$P = -b_0\theta_i$$

But $b_0\theta_i$ is the steady-state value θ_{SS}. Thus

$$\theta_o = \exp(-\zeta\omega_n t)(\theta_{SS}\cos\omega t + Q\sin\omega t) + \theta_{SS} \qquad [32]$$

The overshoot occurs at $\omega t = \pi$, thus

$$\theta_o = \exp(-\zeta\omega_n\pi/\omega)(\theta_{SS} + 0) + \theta_{SS}$$

The overshoot is the difference between the output at that time and the steady-state value. Hence

$$\text{Overshoot} = \theta_{SS}\exp(-\zeta\omega_n\pi/\omega)$$

Since equation [24] gives

$$\omega = \omega_n\sqrt{(1 - \zeta^2)}$$

then

$$\text{Overshoot} = \theta_{SS}\exp\left(\frac{-\zeta\omega_n\pi}{\omega_n\sqrt{(1 - \zeta^2)}}\right)$$

$$\text{Overshoot} = \theta_{SS} \exp\left(\frac{-\zeta\pi}{\sqrt{(1 - \zeta^2)}}\right) \qquad [33]$$

Expressed as a percentage of θ_{SS},

$$\text{Percentage overshoot} = \exp\left(\frac{-\zeta\pi}{\sqrt{(1 - \zeta^2)}}\right) \times 100\% \qquad [34]$$

Table 3.2 gives values of the percentage overshoot for particular damping ratios.

Table 3.2 Percentage peak overshoot

Damping ratio	0.2	0.4	0.6	0.8
Percentage overshoot	52.7	25.4	9.5	1.5

An indication of how fast the oscillations decay is provided by the *subsidence ratio* or *decrement*. This is the amplitude of the second overshoot divided by that of the first overshoot. The first overshoot occurs when $\omega t = \pi$, the second overshoot when $\omega t = 2\pi$. Thus equation [33] gives for the first overshoot

$$\text{First overshoot} = \theta_{SS} \exp\left(\frac{-\zeta\pi}{\sqrt{(1 - \zeta^2)}}\right)$$

and a similar analysis gives for the second overshoot

$$\text{Second overshoot} = \theta_{SS} \exp\left(\frac{-2\zeta\pi}{\sqrt{(1 - \zeta^2)}}\right)$$

Thus the subsidence ratio is

$$\text{Subsidence ratio} = \frac{\text{second overshoot}}{\text{first overshoot}}$$

$$\text{Subsidence ratio} = \frac{\theta_{SS} \exp[(-2\zeta\pi/\sqrt{(1 - \zeta^2)})]}{\theta_{SS} \exp[-\zeta\pi/\sqrt{(1 - \zeta^2)}]}$$

$$\text{Subsidence ratio} = \exp[-\zeta\pi/\sqrt{(1 - \zeta^2)}] \qquad [35]$$

The *settling time* t_s is used as a measure of the time taken for the oscillations to die away. It is the time taken for the response to fall within and remain within some specified percentage, e.g. 2%, of the steady-state value (see Fig. 3.14). This means that the amplitude of the oscillation should be less than 2% of θ_{SS}. Equation [32] indicates how the response θ_o varies with time

$$\theta_o = \exp(-\zeta\omega_n t)(\theta_{SS} \cos\omega t + Q \sin\omega t) + \theta_{SS}$$

The amplitude of the oscillation is $(\theta_o - \theta_{SS})$, thus

$$\text{Amplitude} = \exp(-\zeta\omega_n t)(\theta_{SS} \cos\omega t + Q \sin\omega t)$$

The maximum values of the amplitude occur when ωt is some multiple of π and thus $\cos\omega t = 1$ and $\sin\omega t = 0$. The settling time t_S is when the maximum amplitude is 2% of θ_{SS}, i.e. $0.02\,\theta_{SS}$. Thus

$$0.02\,\theta_{SS} = \exp(-\zeta\omega_n t_s)(\theta_{SS} \times 1 + 0)$$

$$0.02 = \exp(-\zeta\omega_n t_s)$$

Taking logarithms

$$\ln 0.02 = -\zeta\omega_n t_s$$

$\ln 0.02 = -3.9$ or approximately 4. Thus

$$t_s = \frac{4}{\zeta\omega_n} \tag{36}$$

The above is the value of the settling time if the specified percentage is 2%. If the percentage is 5% the equation becomes

$$t_s = \frac{3}{\zeta\omega_n} \tag{37}$$

Since the time taken to complete one cycle, i.e. the so-called *periodic time*, is $1/f$, where f is the frequency, then since $\omega = 2\pi f$ the time to complete one cycle is

$$\text{Periodic time} = \frac{2\pi}{\omega}$$

Hence in a settling time of t_s the number of oscillations that occur is

$$\text{Number of oscillations} = \frac{\text{settling time}}{\text{periodic time}} \tag{38}$$

and thus for a settling time defined for 2% of the steady-state value, combining equation [36] with [38] gives

$$\text{Number of oscillations} = \frac{4/\zeta\omega_n}{2\pi/\omega} = \frac{2\omega}{\pi\zeta\omega_n}$$

Since equation [24] gives

$$\omega = \omega_n\sqrt{(1 - \zeta^2)}$$

then

$$\text{Number of oscillations} = \frac{2\omega_n\sqrt{(1 - \zeta^2)}}{\pi\zeta\omega_n}$$

$$= \frac{2}{\pi}\sqrt{\left(\frac{1}{\zeta^2} - 1\right)} \tag{39}$$

Example 10

A second-order system has a natural angular frequency of 2.0 Hz and a damped frequency of 1.8 Hz. What, for this damping, is (*a*) the damping factor, (b) the 100% rise time, (c) the percentage maximum overshoot, (d) the 2% settling time, and (e) the number of cycles of oscillations that will occur within this settling time?

Answer

(a) Since equation [24] gives

$$\omega = \omega_n \sqrt{(1 - \zeta^2)}$$

$$1.8 = 2.0\sqrt{(1 - \zeta^2)}$$

and so $\zeta = 0.44$.

(b) Since equation [30] gives

$$\omega t_r = \tfrac{1}{2}\pi$$

$$t_r = \frac{\pi}{2\omega} = \frac{\pi}{2 \times 1.8} = 0.87\,\text{s}$$

(c) Since equation [34] gives

$$\text{Percentage overshoot} = \exp\left[\frac{-\zeta\pi}{\sqrt{(1 - \zeta^2)}}\right] \times 100\%$$

$$= \exp\left[\frac{-0.44\,\pi}{\sqrt{(1 - 0.44^2)}}\right] \times 100\%$$

$$= 21\%$$

(d) The 2% settling time is given by equation [36] as

$$t_s = \frac{4}{\zeta\omega_n} = \frac{4}{0.44 \times 2.0} = 4.5\,\text{s}$$

(e) Equation [39] gives, for a 2% settling time,

$$\text{Number of oscillations} = \frac{2}{\pi}\sqrt{\left(\frac{1}{\zeta^2} - 1\right)}$$

$$= \frac{2}{\pi}\sqrt{\left(\frac{1}{0.44^2} - 1\right)} = 1.3$$

Example 11

A TV antenna structure is subject to torsional oscillations when subject to an input torque, such a step input being provided by a sudden gust of wind. The differential equation for the structure is

$$\frac{d^2\theta_o}{dt^2} + 5\frac{d\theta_o}{dt} + 100\,\theta_o = 100\,\theta_i$$

What is the 2% settling time for the structure?

Answer

Comparing the differential equation with equation [17], i.e.

$$\frac{d^2\theta_o}{dt^2} + 2\zeta\omega_n\frac{d\theta_o}{dt} + \omega_n^2\theta_o = b_1\omega_n^2\theta_i$$

then $2\zeta\omega_n = 5$. But the 2% settling time is given by equation [36] as

$$t_s = \frac{4}{\zeta\omega_n} = \frac{4}{5/2} = 1.6\,\text{s}$$

Example 12

A second-order system has an overshoot of 10% and a rise time of 0.4 s when subject to a step input. What is (a) the damping ratio, (b) the damped angular frequency, (c) the undamped angular frequency?

Answer

(a) The percentage overshoot is given by equation [34] as

$$\text{Percentage overshoot} = \exp\left[\frac{-\zeta\pi}{\sqrt{(1 - \zeta^2)}}\right] \times 100\%$$

Thus

$$0.10 = \exp\left[\frac{-\zeta\pi}{\sqrt{(1 - \zeta^2)}}\right]$$

$$\ln 0.10 = \frac{-\zeta\pi}{\sqrt{(1 - \zeta^2)}}$$

$$-2.3\sqrt{(1 - \zeta^2)} = -\zeta\pi$$

$$5.3(1 - \zeta^2) = 9.9\zeta^2$$

Hence $\zeta = 0.6$.

(b) Using equation [30]

$$\omega t_r = \tfrac{1}{2}\pi$$

$$\omega = \frac{\pi}{2 \times 0.4} = 3.9\,\text{Hz}$$

(c) Since equation [24] gives

$$\omega = \omega_n\sqrt{(1 - \zeta^2)}$$

$$\omega_n = \frac{3.9}{\sqrt{(1 - 0.6^2)}} = 4.9\,\text{Hz}$$

D operator and second-order systems

A second-order differential equation [11]

$$a_2\frac{d^2\theta_o}{dt^2} + a_1\frac{d\theta_o}{dt} + a_0\theta_o = b_0\theta_i$$

can be written with D operators as

$$a_2 D^2\theta_o + a_1 D\theta_o + a_0\theta_o = b_0\theta_i$$

$$(a_2 D^2 + a_1 D + a_0)\theta_o = b_0\theta_i$$

$$\frac{\theta_o}{\theta_i} = \frac{b_0}{a_2 D^2 + a_1 D + a_0} \tag{40}$$

The second-order differential equation [17]

$$\frac{d^2\theta_o}{dt^2} + 2\zeta\omega_n\frac{d\theta_o}{dt} + \omega_n^2\theta_o = b_1\omega_n^2\theta_i$$

can be written with D operators as

$$D^2\theta_o + 2\zeta\omega_nD\theta_o + \omega_n^2\theta_o = b_1\omega_n^2\theta_i$$

$$\frac{\theta_o}{\theta_i} = \frac{b_1\omega_n^2}{D^2 + 2\zeta\omega_nD + \omega_n^2} \qquad [41]$$

These equations can be solved by the same approach as adopted for the first-order differential equation.

Example 13

A system can be represented by

$$\frac{\theta_o}{\theta_i} = \frac{1}{(D + 2)(D + 5)}$$

What is the undamped angular frequency and the damping factor for the system?

Answer

The equation can be rearranged to give

$$D^2\theta_o + 7D\theta_o + 10\theta_o = \theta_i$$

This can be compared with equation [17]

$$\frac{d^2\theta_o}{dt^2} + 2\zeta\omega_n\frac{d\theta_o}{dt} + \omega_n^2\theta_o = b_1\omega_n^2\theta_i$$

$$D^2\theta_o + 2\zeta\omega_n^2D\theta_o + \omega_n^2\theta_o = b_1\theta_i$$

Thus $\omega_n^2 = 10$ and so $\omega_n = 3.2\,\text{Hz}$, and $2\zeta\omega_n = 7$ and so $\zeta = 1.1$.

Problems

1 Give examples, with the form of their differential equations, of a first-order system and a second-order system.

2 A first-order system has an output v that varies with time t according to

$$v = V(1 - e^{-3t})$$

where V is the steady-state value of the output. What is the differential equation?

3 A first-order system has a time constant of $4\,\text{s}$ and a steady-state transfer function of 6. What is the form of the differential equation for this system?

4 A mercury-in-glass thermometer has a time constant of $10\,\text{s}$. If it is suddenly taken from being at $20\,°C$ and plunged into hot water at $80\,°C$, what will be the temperature indicated by the thermometer after (*a*) $10\,\text{s}$, (*b*) $20\,\text{s}$?

5 A circuit consists of a resistor R in series with an inductor L. When subject to a step input voltage V at time $t = 0$ the

differential equation for the system is

$$\frac{di}{dt} + \frac{R}{L}i = \frac{V}{L}$$

What is (a) the solution for this differential equation, (b) the time constant, (c) the steady-state current i?

6 What are the time constants and the steady-state transfer functions for the first-order systems described by the following equations?

(a) $0.6 D\theta_o + \theta_o = \theta_i$

(b) $\frac{\theta_o}{\theta_i} = \frac{2}{D + 5}$

(c) $(D + 3)\theta_o = 2\theta_i$

7 Describe the form of the output variation with time for a step input to a second-order system with a damping factor of (a) 0, (b) 0.5, (c) 1.0, (d) 1.5.

8 A *RLC* circuit has a current i which varies with time t when subject to a step input of θ_i according to the following differential equation

$$\frac{d^2i}{dt^2} + 10\frac{di}{dt} + 16i = 16\theta_i$$

What is (a) the undamped frequency, (b) the damping ratio, (c) the solution to the equation if $i = 0$ when $t = 0$ and $di/dt = 0$ when $t = 0$?

9 A system has an output θ_o which varies with time t according to the following differential equation

$$\frac{d^2\theta_o}{dt^2} + 10\frac{d\theta_o}{dt} + 25 = 50\theta_i$$

What is (a) the undamped frequency, (b) the damping ratio, (c) the solution to the equation if $\theta_o = 0$ when $t = 0$ and $d\theta_o/dt = -2$ when $t = 0$ and there is a step input of size 3 units?

10 An accelerometer (an instrument for measuring acceleration) has an undamped angular frequency of 100 Hz and a damping factor of 0.6. What will be (a) the maximum percentage overshoot and (b) the rise time when there is a sudden change in acceleration?

11 What will be (a) the undamped angular frequency, (b) the damping factor, (c) the damped angular frequency, (d) the rise time, (e) the percentage maximum overshoot and (f) the 0.2% settling time for a system which gave the following differential equation?

$$\frac{d^2\theta_o}{dt^2} + 5\frac{d\theta_o}{dt} + 16 = 16\theta_i$$

12 When a voltage of 10 V is suddenly applied to a moving-coil voltmeter it is observed that the pointer of the instrument rises to 11 V before eventually settling down to read 10 V. What is (a) the damping factor and (b) the number of oscillations the pointer will make before it is within 0.2% of the steady value?

4 Laplace transforms

Introduction

The *Laplace transform* is a method of transforming differential equations into more easily solved algebraic equations. This chapter introduces the Laplace transform and considers its use in solving problems which would otherwise require the solution of differential equations.

To help put the concept of mathematical transforms into perspective, a simple example of a mathematical transform is when the problem of multiplication is changed into the simpler operation of addition by means of the *logarithm transform* (Fig. 4.1). Thus the multiplication of B by C to give A,

$$A = BC$$

can be transformed by using logarithms to

$$\log A = \log BC = \log B + \log C$$

We can then add $\log B$ and $\log C$ to give the number D. Then

$$\log A = D$$

To find A we have to carry out the inverse logarithm, or antilogarithm, operation.

$$A = \text{antilog}\, D$$

The *Laplace transform* is a similar type of mathematical operation to this logarithm transform (Fig. 4.2). Differential equations which describe how a circuit behaves with time are transformed into simple algebraic relationships, not involving time, where we can carry out normal algebraic manipulations of the quantities. We talk of the circuit behaviour in the *time domain* being transformed to the *s domain* in which the

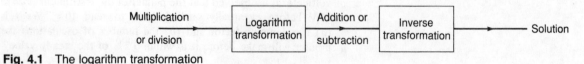

Fig. 4.1 The logarithm transformation

algebraic manipulations can be carried out. Then we use an inverse transform, like the antilogarithm, in order to obtain the solution describing how a signal varies with time, i.e. transform from the *s* domain back to the time domain.

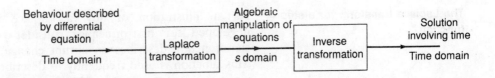

Fig. 4.2 The Laplace transformation

The Laplace transformation

The French mathematician P. S. de Laplace (1749–1827) discovered a means of solving differential equations: multiply each term in the equation by e^{-st} and then integrate each such term with respect to time from zero to infinity; *s* is a constant with the unit of 1/time. The result is what we now call the *Laplace transform*. Thus the Laplace transform of some term which is a function of time is

$$\int_0^\infty (\text{term})\, e^{-st}\, dt$$

Because the term is a function of time it is usually written as $f(t)$ with the Laplace transform, since it is a function of *s*, written as $F(s)$. It is usual to use a capital letter *F* for the Laplace transform and a lower-case letter *f* for the time-varying function $f(t)$. Thus

$$F(s) = \int_0^\infty f(t)\, e^{-st}\, dt \qquad [1]$$

To illustrate the use of the function notation, consider a resistor *R* with the current through it at some instant being *i* and the potential difference across it *v*. Generally, we would write

$$v = Ri$$

But since both *v* and *i* are functions of time, we should ideally indicate this by writing the equation to indicate this, i.e.

$$v(t) = Ri(t)$$

The (t) does not indicate that the preceding term should be multiplied by *t* but just that the preceding term is a function of time, i.e. its value depends on what time we are considering.

If we take the Laplace transforms of *i* and *v* the equation becomes

$$V(s) = RI(s)$$

$V(s)$ indicates that the term is the Laplace transform of $v(t)$, similarly $I(s)$ indicates that the term is the Laplace transform of $i(t)$. The (s) does not indicate that the preceding term should be multiplied by s.

The Laplace transform for a step function

As an illustration of how a Laplace transform can be developed from first principles, consider a step function. The step function describes an abrupt change in some quantity. This function is used frequently to describe the change in the input to a system when a sudden change in its value is made, e.g. the change in the voltage applied to a circuit when it is suddenly switched on. Figure 4.3 shows the form that would be taken by a step input when the abrupt change in input takes place at time $t = 0$ and the size of the step is 1 unit. The equation for this function is

$$f(t) = 1$$

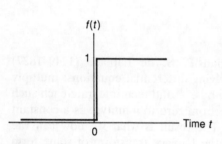

for all values of t greater than 0. For values of t less than 0 the equation is

$$f(t) = 0$$

Fig. 4.3 A step function of height 1

The Laplace transform of this step function, for values greater than 0, is thus

$$F(s) = \int_0^\infty 1 e^{-st}$$

and so

$$F(s) = -\frac{1}{s}\left[e^{-st}\right]_0^\infty$$

Since when $t = \infty$ the value of e^∞ is 0 and when $t = 0$ the value of e^{-0} is -1, then

$$F(s) = \frac{1}{s} \tag{2}$$

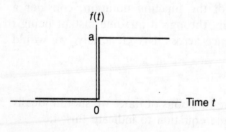

Suppose now instead of a step input signal of height 1 unit we have one of height a units, as in Fig. 4.4. Then, for all values of t greater than 0 we have

$$(f)t = a$$

Fig. 4.4 A step function of height a

The Laplace transform of this function is

$$F(s) = \int_0^\infty a e^{-st}\,dt$$

$$= a \int_0^\infty e^{-st}\,dt$$

But this is just *a* multiplied by the transform of the unit step. Thus

$$F(s) = \frac{a}{s} \tag{3}$$

The multiplication of some function of time by a constant *a* gives a Laplace transform which is just the multiplication of the Laplace transform of that function by the constant.

Example 1

Determine, from first principles, the Laplace transform of the function e^{at}, where *a* is a constant.

Answer

The Laplace transform of this function is found as follows:

$$f(t) = e^{at}$$

$$\text{Laplace transform} = F(s) = \int_0^\infty e^{at} e^{-st} \, dt$$

This can be simplified to

$$F(s) = \int_0^\infty e^{-(s-a)t} \, dt$$

$$F(s) = -\frac{1}{s-a} \left[e^{-(s-a)t} \right]_0^\infty$$

When $t = \infty$ then the term in the brackets becomes 0 and when $t = 0$ it becomes -1. Thus

$$F(s) = \frac{1}{s-a}$$

Using Laplace transforms

Fortunately it is not usually necessary to evaluate the integrals obtained in carrying out the Laplace transformation since tables are available which give the Laplace transforms of all the commonly occurring functions and these combined with some basic rules for handling such transforms enable most problems to be solved.

The basic rules are:

1 The addition of two functions becomes the addition of their two Laplace transforms.

$$f_1(t) + f_2(t) \quad \text{becomes} \quad F_1(s) + F_2(s)$$

2 The subtraction of two functions becomes the subtraction of their two Laplace transforms.

$$f_1(t) - f_2(t) \quad \text{becomes} \quad F_1(s) - F_2(s)$$

3 The multiplication of some function by a constant becomes

the multiplication of the Laplace transform of the function by the same constant.

$$af(t) \quad \text{becomes} \quad aF(s)$$

4 A function which is delayed by a time T, i.e. $f(t - T)$, becomes $e^{-Ts}F(s)$ for values of T greater than or equal to zero.

5 The first derivative of some function becomes s times the Laplace transform of the function minus the value of $f(t)$ at $t = 0$.

$$\frac{\mathrm{d}}{\mathrm{d}t}f(t) \quad \text{becomes} \quad sF(s) - f(0)$$

where $f(0)$ is the value of the function at $t = 0$.

6 The second derivative of some function becomes s^2 times the Laplace transform of the function minus s times the value of the function at $t = 0$ minus the value of the first derivative of $f(t)$ at $t = 0$.

$$\frac{\mathrm{d}^2}{\mathrm{d}t^2}f(t) \quad \text{becomes} \quad s^2F(s) - sf(0) - \frac{\mathrm{d}f(0)}{\mathrm{d}t}$$

where $sf(0)$ is s multiplied by the value of the function at $t = 0$ and $\mathrm{d}f(0)/\mathrm{d}t$ is the first derivative of the function at $t = 0$.

7 The nth derivative of some function becomes s^n times the Laplace transform of the function minus terms involving the values of $f(t)$ and its derivatives at $t = 0$.

$$\frac{\mathrm{d}^n}{\mathrm{d}t^n}f(t) \quad \text{becomes} \quad s^nF(s) - s^{n-1}f(0)$$

$$- \cdots - \frac{\mathrm{d}^{n-1}f(0)}{\mathrm{d}t^{n-1}}$$

8 The first integral of some function, between zero time and time t, becomes $(1/s)$ times the Laplace transform of the function.

$$\int_0^t f(t) \quad \text{becomes} \quad \frac{1}{s}F(s)$$

Table 4.1 gives some of the more common Laplace transforms and their corresponding time functions.

Example 2

Determine, using Table 4.1, the Laplace transforms for:
(a) A step voltage of size $4\,\mathrm{V}$ which starts at $t = 0$.
(b) A step voltage of size $4\,\mathrm{V}$ which starts at $t = 2\,\mathrm{s}$.

Table 4.1 Laplace transforms

Laplace transform	Time function	Description of time function
1		A unit impulse
$\dfrac{1}{s}$		A unit step function
$\dfrac{e^{-st}}{s}$		A delayed unit step function
$\dfrac{1 - e^{-st}}{s}$		A rectangular pulse of duration T
$\dfrac{1}{s^2}$	t	A unit slope ramp function
$\dfrac{1}{s^3}$	$\dfrac{t^2}{2}$	
$\dfrac{1}{s + a}$	e^{-at}	Exponential decay
$\dfrac{1}{(s + a)^2}$	$t\,e^{-at}$	
$\dfrac{2}{(s + a)^3}$	$t^2 e^{-at}$	
$\dfrac{a}{s(s + a)}$	$1 - e^{-at}$	Exponential growth
$\dfrac{a}{s^2(s + a)}$	$t - \dfrac{(1 - e^{-at})}{a}$	
$\dfrac{a}{s(s + a)^2}$	$1 - e^{-at} - ate^{-at}$	
$\dfrac{s}{(s + a)^2}$	$(1 - at)\,e^{-at}$	
$\dfrac{1}{(s + a)(s + b)}$	$\dfrac{e^{-at} - e^{-bt}}{b - a}$	
$\dfrac{ab}{s(s + a)(s + b)}$	$1 - \dfrac{b}{b - a}e^{-at} + \dfrac{a}{b - a}e^{-bt}$	
$\dfrac{1}{(s + a)(s + b)(s + c)}$	$\dfrac{e^{-at}}{(b - a)(c - a)} + \dfrac{e^{-bt}}{(c - a)(a - b)} + \dfrac{e^{-ct}}{(a - c)(b - c)}$	
$\dfrac{\omega}{s^2 + \omega^2}$	$\sin \omega t$	Sine wave
$\dfrac{s}{s^2 + \omega^2}$	$\cos \omega t$	Cosine wave
$\dfrac{\omega}{(s + a)^2 + \omega^2}$	$e^{-at} \sin \omega t$	Damped sine wave
$\dfrac{s + a}{(s + a)^2 + \omega^2}$	$e^{-at} \cos \omega t$	Damped cosine wave
$\dfrac{\omega^2}{s(s^2 + \omega^2)}$	$1 - \cos \omega t$	
$\dfrac{\omega^2}{s^2 + 2\zeta\omega s + \omega^2}$	$\dfrac{\omega}{\sqrt{(1 - \zeta^2)}}e^{-\zeta\omega t} \sin\left[\omega\sqrt{(1 - \zeta^2)}t\right]$	
$\dfrac{\omega^2}{s(s^2 + 2\zeta\omega s + \omega^2)}$	$1 - \dfrac{1}{\sqrt{(1 - \zeta^2)}}e^{-\zeta\omega t} \sin\left[\omega\sqrt{(1 - \zeta^2)}t + \phi\right]$	
with $\zeta < 1$	with $\zeta = \cos \phi$	

(*c*) A ramp voltage which starts at $t = 0$ and increases at the rate of 3 V/s.

(*d*) A ramp voltage which starts at $t = 2$ s and increases at the rate of 3 V/s.

(*e*) An impulse voltage of size 4 V which starts at $t = 3$ s.

(*f*) A sinusoidal voltage of amplitude 2 V and angular frequency 10 Hz.

Answer

Figure 4.5 shows the form of the six functions, the six representing common forms of signals input to systems.

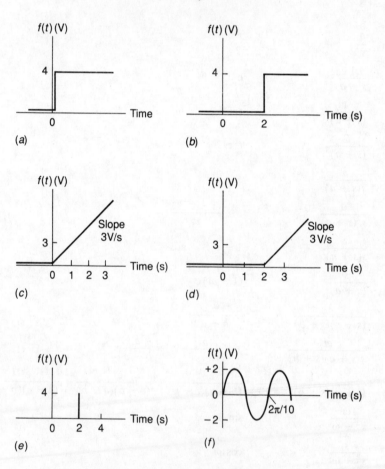

Fig. 4.5 (*a*) Step function, (*b*) delayed step function, (*c*) ramp function, (*d*) delayed ramp function, (*e*) delayed impulse, (*f*) sinusoidal function

(*a*) The step voltage is a function of the form

$$f(t) = a$$

where *a* has the value, in this case, of 4 V. The Laplace transform of the step function of size 1 is $1/s$ and thus the step function of size *a* has the the Laplace transform of

$$F(s) = a \times \frac{1}{s}$$

Hence

$$F(s) = \frac{4}{s}$$

(b) The step function in (a) is delayed by 2 s. For a delayed function the Laplace transform is that of the undelayed function, i.e. the function starting at $t = 0$, multiplied by e^{-sT}. Thus the Laplace transform is

$$F(s) = \frac{a}{s} e^{-sT} = \frac{4}{s} e^{-2s}$$

(c) The ramp function is of the form

$$f(t) = at$$

with a having the value 3 V/s. Because a is a constant then the Laplace transform of the function will be a multiplied by the transform of t which is $1/s^2$. Thus

$$F(s) = \frac{a}{s^2} = \frac{3}{s^2}$$

(d) The ramp voltage is delayed by a time T, where $T = 3$ s. For a delayed function the Laplace transform is that of the undelayed function, i.e. the function starting at $t = 0$, multiplied by e^{-sT}. Thus the Laplace transform is

$$F(s) = \frac{a e^{-T/s}}{s^2} = \frac{3 e^{-2s}}{s^2}$$

(e) The Laplace transform of a unit impulse occurring at $t = 0$ is 1. For an impulse of 4 V the transform will be 4. Delaying the impulse means the undelayed function is multiplied by $e^{-T/s}$. Thus the Laplace transform with $T = 3$ s is

$$F(s) = 4 e^{-3/s}$$

(f) The Laplace transform of a sinusoidal function $\sin \omega t$ is

$$F(s) = \frac{\omega}{s^2 + \omega^2}$$

Thus the transform for a sinusoidal function of amplitude A, i.e. the function $A \sin \omega t$, is

$$F(s) = \frac{A\omega}{s^2 + \omega^2}$$

Thus for amplitude 2 V and angular frequency 10 Hz,

$$F(s) = \frac{20}{s^2 + 100}$$

Example 3

Determine, using Table 4.1, the Laplace transforms for the following functions:

(a) t^2
(b) $t^2 e^{-at}$
(c) $t^2 (1 + e^{-at})$

Answer

(a) The table gives the Laplace transform of $\frac{1}{2}t^2$ as $1/s^3$. Thus to obtain the Laplace transform of t^2 we have to multiply the function in the table by 2. Since this is a constant, the Laplace transform of t^2 will be

$$F(s) = \frac{2}{s^3}$$

(b) Using the table, the transform is

$$F(s) = \frac{2}{(s + a)^3}$$

Note that the Laplace transform of two functions multiplied together is *not* the multiplication of their separate Laplace transforms.

(c) The Laplace transform of two functions added together is the addition of their separate Laplace transforms.

$$f(t) = t^2 + t^2 e^{-at}$$

$$F(s) = \frac{2}{s^3} + \frac{2}{(s + a)^3}$$

Example 4

Determine, using Table 4.2, the inverse transformations of:

(a) $\dfrac{2}{s}$

(b) $\dfrac{3}{2s + 1}$

(c) $\dfrac{2}{s - 5}$

Answer

To use Table 4.1 to obtain the inverse transformations means looking through the table to find the Laplace transform which is of the same basic form.

(a) The table includes a Laplace transform of $1/s$ and thus since this is just multiplied by the constant 2 the inverse transformation will be the function which gives $1/s$, i.e. 1 multiplied by the same constant. The inverse transformation is thus 2.

(b) This transform can be rearranged to give

$$\frac{(3/2)}{s + (1/2)}$$

The table contains the transform $1/(s + a)$, the inverse of which is e^{-at}. Thus the inverse transformation is just this multiplied by the constant $(3/2)$ with $a = (1/2)$, i.e. $(3/2)e^{-t/2}$.

(c) This transform is of the same form as in (b) with $a = -5$. Thus the inverse transformation is $2e^{5t}$.

Using Laplace transforms to solve differential equations

In using Laplace transforms to obtain the solution to a differential equation, the procedure adopted is:

1 Transform each term in the differential equation into its Laplace transform, i.e. change the function of time into a function of (s).
2 Carry out all algebraic manipulations, e.g. considering what happens when a step input is applied to the system.
3 Convert the resulting Laplace function back into an equation giving a function of time, i.e. invert the Laplace transformation operation. In order to use tables of Laplace transforms to carry out the conversion, it is often necessary first to use partial fractions to get it into the standard forms given in tables (see later this chapter).

Example 5

Use Laplace transforms to solve the following differential equation:

$$3\frac{dx}{dt} + 2x = 4$$

with $x = 0$ at $t = 0$.

Answer

The Laplace transform of $3\,dx/dt$ is 3 times the Laplace transform of dx/dt. The Laplace transform of $2x$ is 2 times the Laplace transform of x. The Laplace transform of 4 is, since this can be considered to be a step function of height 4, $4/s$. Thus

$$3[sX(s) - x(0)] + 2X(s) = 4/s$$

where $X(s)$ is the Laplace transform of x. Since $x(0) = 0$ then

$$3[sX(s) - 0] + 2X(s) = 4/s$$

and so

$$3s^2X(s) + 2sX(s) = 4$$

$$X(s) = \frac{4}{3s^2 + 2s} = \frac{2(2/3)}{s[s + (2/3)]}$$

We now need to find the functions which would give the Laplace transforms of this form in order to obtain the inverse transformation and obtain x. Since the inverse transformation of $a/[s(s + a)]$ is $(1 - e^{-at})$ then

$$x = 2(1 - e^{-2t/3})$$

Example 6

For a voltage step input of size V at $t = 0$ into a series CR circuit the differential equation for the potential difference across the capacitor v_C is given by

$$V = RC\frac{dv_C}{dt} + v_C$$

v_C is zero at $t = 0$. Use Laplace transforms to solve this equation.

Answer

The Laplace transform of a unit step input is $1/s$ and thus for one of size V it is V/s. The Laplace transform for dv_C/dt is $[sV_C(s) - 0]$, since the function v_C is zero at $t = 0$. Thus the Laplace transform for $RCdv_C/dt$ is $RCsV_C(s)$. The Laplace transform of v_C is $V_C(s)$. Thus the transform of the differential equation is

$$\frac{V}{s} = RCsV_C(s) + V_C(s)$$

Thus

$$V_C(s) = \frac{V}{(RCs + 1)s}$$

This can be rearranged to give

$$V_C(s) = \frac{V(1/RC)}{[s + (1/RC)]s}$$

The function $(1 - e^{-at})$ gives the Laplace transform

$$\frac{a}{(s + a)s}$$

Thus, with $a = (1/RC)$,

$$v_C = V(1 - e^{-t/RC}).$$

Figure 4.6 shows a graph of this equation. Such a form of equation and graph is typical of first-order systems subject to a step input. RC is the time constant τ (see Chapter 3).

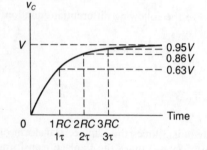

v_C

V ---- 0.95V
---- 0.86V
---- 0.63V

Time

0 1RC 2RC 3RC
   1τ 2τ 3τ

Fig. 4.6 Example 6

Example 7

For a step input at $t = 0$ of size V into a series LR circuit the current variation with time is described by the equation

$$\frac{L}{R}\frac{di}{dt} + i = \frac{V}{R}$$

The current i is zero at $t = 0$. Using Laplace transforms solve this equation.

Answer

The Laplace transform of di/dt is $sI(s)$, since $i(0)$ is zero, and so for $(L/R)di/dt$ it is $(L/R)sI(s)$. The Laplace transform for i is $I(s)$. The Laplace transform for a unit step input is $1/s$ and thus for one of size (V/R) it is $(V/R)/s$. Hence, the transform of the differential equation can be written as

$$(L/R)sI(s) + I(s) = \frac{(V/R)}{s}$$

Hence

$$I(s) = \frac{(V/R)}{[(L/R)s + 1]s}$$

This can be rearranged to give

$$I(s) = \frac{(V/R)(R/L)}{[s + (R/L)]s}$$

The function $(1 - e^{-at})$ gives the Laplace transform

$$\frac{a}{(s + a)s}$$

Thus, with $a = (R/L)$,

$$i = (V/R)(1 - e^{-Rt/L})$$

The graph of this equation is of a similar form to that given in Fig. 4.6, the time constant τ being L/R and the current reached eventually being V/R.

Example 8

When a mercury-in-glass thermometer is inserted into a hot liquid there is essentially a step input of temperature θ_i to the thermometer, where θ_i is the temperature of the hot liquid. The relationship between the output of the thermometer θ_o, i.e. its reading, and time is given by the first-order differential equation

$$K\frac{d\theta_o}{dt} = \theta_i - \theta_o$$

where θ_o is a function of time. Use the Laplace transformation to obtain a solution for this equation. Take the value of θ_o to be 0 at $t = 0$, i.e. in this problem we are only concerned with the change in temperature when the thermometer is inserted into the liquid. Thus θ_i is the temperature of the hot liquid relative to that of the thermometer before it is inserted into the liquid.

Answer

The Laplace transform of $d\theta_o/dt$ is $[s\theta_o(s) - 0]$, since the value of θ_o is 0 at $t = 0$. Thus the transform for $Kd\theta_o/dt$ is $Ks\theta_o(s)$. The transform of θ_i is $(1/s)\theta_i$, since it is a step input, and that of θ_o is $\theta_o(s)$. Hence the transform of the differential equation is

$$Ks\theta_o(s) = (1/s)\theta_i - \theta_o(s)$$

Hence

$$\frac{\theta_o(s)}{\theta_i} = \frac{1}{s(Ks + 1)}$$

This can be rearranged to give

$$\frac{\theta_o(s)}{\theta_i} = \frac{(1/K)}{s[s + (1/K)]}$$

The function $(1 - e^{-at})$ gives the Laplace transform

$$\frac{a}{(s + a)s}$$

Thus

$$\theta_o = \theta_i(1 - e^{-t/K})$$

The graph of this equation is of a similar form to that in Fig. 4.6, the time constant τ being K and the temperature reached eventually being θ_i.

Partial fractions

The process of converting an algebraic expression into simple fraction terms is called resolving into *partial fractions*. For example, it is the conversion of

$$\frac{3x + 4}{x^2 + 3x + 2} \quad \text{into} \quad \frac{1}{x + 1} + \frac{2}{x + 2}$$

In order to resolve an algebraic expression into partial fractions the denominator must factorize, e.g. in the above expression the $(x^2 + 3x + 2)$ yields the factors $(x + 1)$ and $(x + 2)$, and the numerator must be at least one degree less than the denominator, e.g. in the above example the numerator only contains x to the power 1 while the denominator has x to the power 2. When the degree of the numerator is equal to or higher than the degree of the denominator, the numerator must be divided by the denominator to give terms which each have the numerator at least one degree less than the denominator.

There are basically three types of *partial fractions*. The form of the partial fractions for each of these types is as follows:

1 Linear factors in the denominator

Expression $\dfrac{f(s)}{(s + a)(s + b)(s + c)}$

Partial fraction $\dfrac{A}{s + a} + \dfrac{B}{s + b} + \dfrac{C}{s + c}$

2 Repeated linear factors in the denominator

Expression $\dfrac{f(s)}{(s + a)^n}$

Partial fraction $\dfrac{A}{s + a} + \dfrac{B}{(s + a)^2} + \dfrac{C}{(s + a)^3}$

$$+ \ldots \frac{N}{(s + a)^n}$$

3 Quadratic factors in the denominator, when the quadratic does not factorize without imaginary terms

Expression $\dfrac{f(s)}{as^2 + bs + c}$

Partial fraction $\dfrac{As + B}{as^2 + bs + c}$

or if there is also a linear factor in the denominator,

$$\text{Expression } \frac{f(t)}{(as^2 + bs + c)(s + d)}$$

$$\text{Partial fraction } \frac{As + B}{as^2 + bs + c} + \frac{C}{s + d}$$

Whatever the form of the partial fractions, the values of the constants A, B, C, etc., can be found by combining the partial fractions into an expression which has the same denominator as the original equation. Thus if the partial fractions for

$$\frac{3x + 4}{(x + 1)(x + 2)} \quad \text{are} \quad \frac{A}{x + 1} + \frac{B}{x + 2}$$

Then

$$\frac{3x + 4}{(x + 1)(x + 2)} = \frac{A(x + 2) + B(x + 1)}{(x + 1)(x + 2)}$$

Then the numerator of the partial fraction expression must have the same value as that of the original equation.

$$3x + 4 = A(x + 2) + B(x + 1)$$

This must be true for all values of the variable. The procedure is then to pick values of the variable which enable some of the terms involving constants to become zero and so enable other constants to be determined. Thus if we let $x = -2$ then

$$3(-2) + 4 = A(-2 + 2) + B(-2 + 1)$$

and so $B = 2$. If we let $x = -1$, then

$$3(-1) + 4 = A(-1 + 2) + B(-1 + 1)$$

and so $A = 1$. Thus

$$\frac{3x + 4}{(x + 1)(x + 2)} = \frac{1}{x + 1} + \frac{2}{x + 2}$$

Example 9

Determine the partial fractions of

$$\frac{s + 5}{s^2 + 3s + 2}$$

Answer

The expression can be rearranged to give

$$\frac{s + 5}{(s + 1)(s + 2)}$$

This can be resolved into the partial fractions

$$\frac{A}{s+1} + \frac{B}{s+2}$$

This is equal to

$$\frac{A(s+2) + B(s+1)}{(s+1)(s+2)}$$

Since this expression has the same denominator as the initial expression then we must have, for the expressions to be equal,

$$A(s+2) + B(s+1) = s+5$$

To determine the values of the constants A and B, values of s are chosen to make the terms in A or B equal to zero. Thus if we choose $s = -1$, then

$$A(-1+2) + B(-1+1) = -1+5$$

and A equals 4. When $s = -2$, then

$$A(-2+2) + B(-2+1) = -2+5$$

and B equals -3. Thus the expression in partial fractions is

$$\frac{4}{s+1} + \frac{-3}{s+2}$$

Example 10

Consider a series CR circuit which has a ramp voltage input. The differential equation for the potential difference across the capacitor v_C in such a circuit is

$$RC\frac{dv_C}{dt} + v_C = Vt$$

Vt is the ramp input voltage, with V being the voltage rise per second. When $t = 0$ the value of v_C is 0.

Answer

Note that this equation differs from the differential equation for the step input voltage V in Example 7 by having Vt, with V being a rate of change of voltage rather than the steady value of the step.

The Laplace transform for dv_C/dt is $sV_C(s)$ and so $RCdv_C/dt$ is $RCsV_C(s)$. The transform for v_C is $V_C(s)$. The transform for t is $1/s^2$ and so for Vt it is V/s^2. Thus the transform for the differential equation is

$$RCsV_C(s) + V_C(s) = \frac{V}{s^2}$$

Hence

$$V_C(s) = \frac{V}{(RCs+1)s^2}$$

This can be rearranged to give

$$V_C(s) = \frac{V(1/RC)}{[s + (1/RC)]s^2}$$

Partial fractions have to be used to obtain an expression in a suitable form for the inverse transformation to be made. Since the denominator involves s^2 the quadratic form of partial fractions has to be used. Thus

$$\frac{a}{(s+a)s^2} = \frac{A}{s+a} + \frac{(Bs+D)}{s^2} = \frac{As^2 + (Bs+D)(s+a)}{(s+a)s^2}$$

where $a = 1/RC$, and A, B and D are constants. Thus, since the numerators must be equal,

$$a = As^2 + Bs^2 + Bsa + Ds + Da$$

If we let $s = -a$, then

$$a = Aa^2 + Ba^2 - Ba^2 - Da + Da$$

and so $A = 1/a$. If we let $s = 0$, then

$$a = 0 + 0 + 0 + 0 + Da$$

and so $D = 1$. If we let $s = +a$, then

$$a = Aa^2 + Ba^2 + Ba^2 + Da + Da$$

and since $A = 1/a$ and $D = 1$, then

$$a = a + Ba^2 + Ba^2 + a + a$$

and so $B = -1/a$. Thus the expression can be written as

$$\frac{1/a}{(s+a)} + \frac{[(-1/a)s + 1]}{s^2} = \frac{1/a}{(s+a)} - \frac{(1/a)}{s} + \frac{1}{s^2}$$

The function e^{-at} has the Laplace transform $1/(s+a)$, the unit step function the transform $1/s$ and the ramp function t the transform $1/s^2$. Hence

$$v_C = V[(RC)e^{-t/RC} - (RC) + t]$$

and so

$$v_C = Vt - VRC(1 - e^{-t/RC})$$

Figure 4.7 shows the graph of this equation. The easiest way to imagine this graph is the straight line graph $v_C = Vt$ minus the graph $v_C = VRC(1 - e^{-t/RC})$. The time constant is RC.

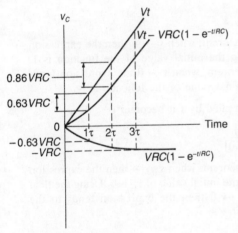

v_C

Vt

$Vt - VRC(1 - e^{-t/RC})$

$0.86\,VRC$

$0.63\,VRC$

0 1τ 2τ 3τ Time

$-0.63\,VRC$

$-VRC$

$VRC(1 - e^{-t/RC})$

Fig. 4.7 Example 10

The initial and final value theorems

If a Laplace transform is multiplied by s, the value of the product as s tends to infinity is the value of the inverse transform as the time t tends to zero.

$$\lim_{s \to \infty} sF(s) = \lim_{t \to 0} f(t) \tag{4}$$

This is known as the *initial value theorem*.

If a Laplace transform is multiplied by s, the value of the product as s tends to zero is the value of the inverse transform as t tends to zero.

$$\text{limit } sF(s) = \text{limit } f(t) \tag{5}$$
$$\text{\tiny } s\to0 \qquad\qquad t\to\infty$$

This is known as the *final value theorem*.

The initial and the final value theorems are useful when there is a need to determine from the Laplace transform the behaviour of the function $f(t)$ at 0 and ∞.

Example 11

Without evaluating the Laplace transforms, what are the initial and final values of the functions giving the following transforms?

(a) $\quad F(s) = \dfrac{s + a}{s^2}$

(b) $\quad V_C(s) = \dfrac{V(1/RC)}{[s + (1/RC)]s}$

Answer

(a) If the expression is multiplied by s it becomes

$$sF(s) = \frac{s + a}{s} = 1 + \frac{a}{s}$$

Using the initial value theorem, when $s \to \infty$ then the expression tends to the value 1. So the initial value of the function is 1. Using the final value theorem, when $s \to 0$ the expression tends to the value ∞. So the final value of the function is ∞.

(b) If the expression is multiplied by s it becomes

$$sV_C(s) = \frac{V(1/RC)}{[s + (1/RC)]}$$

Using the initial value theorem, when $s \to \infty$ then the expression tends to the value 0. So the initial value of v_C is 0. Using the final value theorem, when $s \to 0$ then the expression tends to the value $V(1/RC)/(1/RC)$ or V.

Problems

1   Determine, using Table 4.1, the Laplace transforms for:

(a) A step voltage of size 6 V which starts at $t = 0$ s.
(b) A step voltage of size 6 V which starts at $t = 3$ s.
(c) A ramp voltage of 6 V/s which starts at $t = 0$ s.
(d) A ramp voltage of 6 V/s which starts at $t = 3$ s.
(e) An impulse of size 6 V at $t = 0$ s.
(f) An impulse of size 6 V at $t = 3$ s.
(g) A sinusoidal signal of amplitude 6 V and frequency 50 Hz which starts at $t = 0$ s.

2   Determine, using Table 4.1, the Laplace transforms for:

(a) e^{-2t}
(b) $5e^{-2t}$
(c) $V_0 e^{-t/\tau}$

(d) $1 - e^{-2t}$
(e) $5(1 - e^{-2t})$
(f) $V_0(1 - e^{-t/\tau})$

3   Determine, using Table 4.1, the inverse of the following Laplace transforms:

(a) $2/(s + 3)$
(b) $2/(3s + 1)$
(c) $2/s(s + 3)$
(d) $2/s(3s + 1)$

4   Solve the following differential equations:

(a) $2\dfrac{dx}{dt} + 5x = 6$ with $x = 0$ at $t = 0$

(b) $8\dfrac{dx}{dt} + x = 4$ with $x = 0$ at $t = 0$

5   Determine the Laplace transforms of the following voltages which vary with time according to the given equations:
(a) $v = 5(1 - e^{-t/50})$
(b) $v = 10 + 5(1 - e^{-t/50})$
(c) $v = 5e^{-t/50}$

6   Determine the Laplace transforms of the voltages given by the following differential equations:

(a) $\dfrac{dv}{dt} + 2v = 0$ with $v = 0$ when $t = 0$

(b) $\dfrac{dv}{dt} + 2v = 9$ with $v = 0$ when $t = 0$

7   Using the final and initial value theorems, what are the final and initial values of the signals which gave the following Laplace transforms?

(a) $\dfrac{5}{s}$

(b) $\dfrac{5}{s(s + 2)}$

8   Using partial fractions, determine the signal variation with time which gave the following Laplace transforms:

(a) $\dfrac{4s - 5}{s^2 - s - 2}$

(b) $\dfrac{6s + 8}{s(s + 1)(s + 2)}$

(c) $\dfrac{1}{s^2 + 3s + 2}$

9   Solve, using Laplace transforms, the following second-order differential equations:

(a) $\dfrac{d^2x}{dt^2} + 64x = 0$ with $\dfrac{dx}{dt} = 0$ and $x = 2$ when $t = 0$

(b) $\dfrac{d^2x}{dt^2} + 64x = 0$ with $\dfrac{dx}{dt} = 2$ and $x = 0$ when $t = 0$

5 Dynamic system models

In Chapter 1 control systems were considered for the steady-state situation, i.e. the transfer function was not assumed to change with time. No account was taken of how the output of such a system would vary with time when there was an input. Chapter 2 in looking at models for the basic passive building blocks of systems has however led to input-output relationships which are time dependent and involve differential equations. Chapters 3 and 4 were a consideration of the mathematics of solving such differential equations. This chapter brings together all these earlier chapters in a consideration of the behaviour of control systems when changes with time are not ignored, i.e. a consideration of the dynamic behaviour of systems.

Transfer functions of dynamic elements

Suppose we have a system where the input θ_i is related to the output θ_o by the differential equation

$$a_2 \frac{d^2\theta_o}{dt^2} + a_1 \frac{d\theta_o}{dt} + a_0\theta_o = b_1\theta_i$$

where a_2, a_1, a_0 and b_1 are constants. If all the initial conditions are zero then the Laplace transform of this equation is

$$a_2 s^2 \theta_o(s) + a_1 s \theta_o(s) + a_0 \theta_o(s) = b_1 \theta_i(s)$$

$$\frac{\theta_o(s)}{\theta_i(s)} = \frac{b_1}{a_2 s^2 + a_1 s + a_0}$$

The *transfer function* $G(s)$ of a linear system which describes the dynamic behaviour is defined as the ratio of the Laplace transform of the output variable $\theta_o(s)$ to the Laplace transform of the input variable $\theta_i(s)$, with all initial conditions assumed to be zero. See Chapter 2 for an explanation of the term linear.

$$G(s) = \frac{\theta_o(s)}{\theta_i(s)} \qquad\qquad [1]$$

Hence, for the system giving the above equation

$$G(s) = \frac{\theta_o(s)}{\theta_i(s)} = \frac{b_1}{a_2 s^2 + a_1 s + a_0}$$

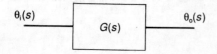

Thus if we represent a system by a block diagram, then $G(s)$ is the 'function' in the box which takes an input of $\theta_i(s)$ and gives an output $\theta_o(s)$ (Fig. 5.1).

Fig. 5.1 Block diagram representation

Example 1

Write down the transfer function $G(s)$ for systems giving the following input–output relationships:

(a) A spring-dashpot-mass system (Chapter 2 equation [13]), with input F and output x

$$m\frac{d^2 x}{dt^2} + c\frac{dx}{dt} + kx = F$$

(b) A resistor-capacitor circuit (Chapter 2 equation [26]), with input v and output v_C

$$v = RC\frac{dv_C}{dt} + v_C$$

(c) A resistor-capacitor-inductor circuit (Chapter 2 equation [27]), with input v and output v_C

$$v = RC\frac{dv_C}{dt} + LC\frac{d^2 v_C}{dt^2} + v_C$$

(d) An electrical system (Chapter 2 equation [30], with input v and output v_C

$$v = \frac{R}{L}\int v_C\,dt + RC\frac{dv_C}{dt} + v_C$$

(e) A hydraulic system (Chapter 2 equation [51]), with input q_1 and output h

$$q_1 = A\frac{dh}{dt} + \frac{\pi g h}{R}$$

(f) The elements in the armature controlled d.c. motor system (Chapter 2 Fig. 2.39(a)):

Armature circuit: input $(v_a - v_b)$, output i_a

$$v_a - v_b = L_a\frac{di_a}{dt} + R_a i_a$$

Armature coil: input i_a, output T

$$T = k_4 i_a$$

Load: input T output ω

$$I\frac{d\omega}{dt} = T - \omega$$

Feedback loop: input ω, output v_b

$$v_b = k_3\omega$$

(g) Hydraulic system with load (Chapter 2 equation [90])

$$\tau\frac{d^2x_o}{dt^2} + \frac{dx_o}{dt} = kx_i$$

Answer

(a) Carrying out the Laplace transform of the equation with all initial conditions zero gives

$$ms^2X(s) + csX(s) + kX(s) = F(s)$$

Hence

$$G(s) = \frac{X(s)}{F(s)} = \frac{1}{ms^2 + cs + k}$$

(b) Carrying out the Laplace transform of the equation with all initial conditions zero gives

$$RCsV_C(s) + V_C(s) = V(s)$$

Hence

$$G(s) = \frac{V_C(s)}{V(s)} = \frac{1}{RCs + 1}$$

(c) Carrying out the Laplace transform of the equation with all initial conditions zero gives

$$RCsV_C(s) + LCs^2V_C(s) + V_C(s) = V(s)$$

Hence

$$G(s) = \frac{V_C(s)}{V(s)} = \frac{1}{LCs^2 + RCs + 1}$$

(d) Carrying out the Laplace transform of the equation with all initial conditions zero gives

$$\frac{R}{L}\frac{1}{s}V_C(s) + RCsV_C(s) + V_C(s) = V(s)$$

Hence

$$G(s) = \frac{V_C(s)}{V(s)} = \frac{1}{(R/L)(1/s) + RCs + 1}$$

$$= \frac{s}{(R/L) + RCs^2 + s}$$

(e) Carrying out the Laplace transform of the equation with all initial conditions zero gives

$$AsH(s) + (\rho g/R)H(s) = Q_1(s)$$

Hence

$$G(s) = \frac{H(s)}{Q_1(s)} = \frac{1}{As + (\rho g/R)}$$

(f) Carrying out the Laplace transform of the equation with all initial conditions zero gives

Armature circuit

$$(V_a - V_b)(s) = L_a s I_a(s) + R_a I_a(s)$$

$$G(s) = \frac{I_a(s)}{(V_a - V_b)(s)} = \frac{1}{L_a s + R_a}$$

Armature coil

$$G(s) = \frac{T(s)}{i_a(s)} = k_4$$

Load

$$I s \omega(s) = T(s) - c\omega(s)$$

$$G(s) = \frac{\omega(s)}{T(s)} = \frac{1}{I s + c}$$

Feedback

$$G(s) = \frac{V_b(s)}{\omega(s)} = k_3$$

(g) Carrying out the Laplace transform of the equation with all initial conditions zero gives

$$\tau s^2 X_o(s) + s X_o(s) = k X_i(s)$$

$$G(s) = \frac{X_o(s)}{X_i(s)} = \frac{k}{s(\tau s + 1)}$$

First-order and second-order elements

The *order* of an element, or system, can be defined as being the highest power of derivative in the differential equation. Alternatively, the *order* of an element, or system, can be defined as being the highest power of s in the denominator. Thus a first-order element will only have s to the power 1 in the denominator, while a second-order element will have the highest power of s in the denominator being two.

For a *first-order element* the differential equation is of the form

$$a_1 \frac{d\theta_o}{dt} + a_o \theta_o = b_o \theta_i \qquad [2]$$

The corresponding Laplace transform of the equation is, if $\theta_o = 0$ at $t = 0$,

$$a_1 s \times \theta_o(s) + a_o \times \theta_o(s) = b_o \times \theta_i(s)$$

Hence

$$G(s) = \frac{b_o}{a_1 s + a_o}$$

This can be rearranged to give

$$G(s) = \frac{b_o/a_o}{(a_1/a_o)s + 1} \tag{3}$$

b_o/a_o is the steady-state transfer function G of the system. a_1/a_o is the time constant τ of the system. Hence

$$G(s) = \frac{G}{\tau s + 1} \tag{4}$$

This is the general form that is taken for the output-input relationship in the s domain for a first-order system.

The relationship between the input θ_i and output θ_o for a *second-order* element is described by the differential equation (see Chapter 3)

$$a_2 \frac{d^2\theta_o}{dt^2} + a_1 \frac{d\theta_o}{dt} + a_0\theta_o = b_0\theta_i \tag{5}$$

where b_0, a_0, a_1 and a_2 are constants. If at $t = 0$ we have $\theta_o = 0$ and $d\theta_o/dt = 0$ then the Laplace transform is

$$a_2s^2 \times \theta_o(s) + a_1s \times \theta_o(s) + a_0 \times \theta_o(s) = b_0 \times \theta_i(s)$$

Hence

$$G(s) = \frac{\theta_o(s)}{\theta_i(s)} = \frac{b_0}{a_2s^2 + a_1s + a_0}$$

This can be rearranged to give

$$G(s) = \frac{(b_0/a_0)}{(a_2/a_0)s^2 + (a_1/a_0)s + 1} \tag{6}$$

As indicated in Chapter 3, the second-order differential equation can be written in terms of the natural frequency ω_n and the damping ratio ζ.

$$\frac{d^2\theta_o}{dt^2} + 2\zeta\omega_n \frac{d\theta_o}{dt} + \omega_n^2\theta_o = b_0\omega_n^2\theta_i \tag{7}$$

where ω_n is the angular frequency with which the system will freely oscillate in the absence of any damping and ζ is the *damping ratio*. The Laplace transform is

$$s^2\theta_o(s) + 2\zeta\omega_ns\theta_o(s) + \omega_n^2\theta_o(s) = b_0\omega_n^2\theta_i(s)$$

Hence

$$G(s) = \frac{\theta_o(s)}{\theta_i(s)} = \frac{b_0\omega_n^2}{s^2 + 2\zeta\omega_ns + \omega_n^2} \tag{8}$$

This is the general form taken by a second-order system in the s domain.

Example 2

What are the orders of the elements described by the transfer function answers in Example 1?

(a) $G(s) = \dfrac{1}{ms^2 + cs + k}$

(b) $G(s) = \dfrac{1}{RCs + 1}$

(c) $G(s) = \dfrac{1}{LCs^2 + RCs + 1}$

(d) $G(s) = \dfrac{s}{(R/L) + RCs^2 + s}$

(e) $G(s) = \dfrac{1}{As + (\varrho g/R)}$

Answer

(a) Second order since the denominator has the highest s term of s^2.
(b) First order since the denominator has the highest s term of s.
(c) Second order since the denominator has the highest s term of s^2.
(d) Second order since the denominator has the highest s term of s^2.
(e) First order since the denominator has the highest s term of s.

Step input to a first-order system

Consider the behaviour of a first-order system when subject to a step input. For a first-order system we can take the relationship to be of the form given by equation [4], namely

$$G(s) = \frac{G}{\tau s + 1}$$

The Laplace transform of the output is thus

$$G(s) \times \text{Laplace transform of input}$$

$$\frac{G}{\tau s + 1} \times \text{Laplace transform of input}$$

The Laplace transform for a one unit step input at $t = 0$ is $1/s$. Hence, for such an input

$$\text{Laplace transform of output} = \frac{G}{\tau s + 1} \times \frac{1}{s}$$

$$= G \times \frac{(1/\tau)}{s[s + (1/\tau)]}$$

The transform is of the form

$$\frac{a}{s(s + a)}$$

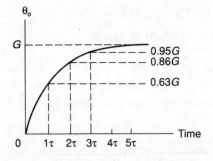

Fig. 5.2 $\theta_o = G[1 - e^{-t/\tau}]$ for a unit step input

where $a = (1/\tau)$. Hence, for a unit step input,

$$\theta_o = G[1 - e^{-t/\tau}] \qquad [9]$$

Figure 5.2 is a graph of this equation. If the step has a size A then

$$\theta_o = AG[1 - e^{-t/\tau}] \qquad [10]$$

Example 3

A thermocouple has the transfer function linking its output in volts to its input θ_i in °C of

$$G(s) = \frac{30 \times 10^{-6}}{10s + 1}$$

What will be (a) the time taken for the output of the thermocouple to reach 95% of its final value and (b) the final steady value when there is a step input of 100 °C?

Answer

(a) As the transfer function indicates, the thermocouple is a first-order system. Thus comparing the transfer function with equation [4], i.e.

$$G(s) = \frac{G}{\tau s + 1}$$

$G = 30 \times 10^{-6}$ V/°C and $\tau = 10$ s. The time taken to reach 95% of the output is 3τ (as in Fig. 4.8) and thus 30 s.

(b) We can use the final value theorem (equation [5], Chapter 4)

$$\lim_{s \to 0} sF(s) = \lim_{t \to \infty} f(t)$$

For a step input of size θ_i the Laplace transform of the output is θ_i/s, hence

$$\text{Laplace transform of output} = \frac{G}{\tau s + 1} \times \frac{\theta_i}{s}$$

Thus

$$sF(s) = \frac{G\theta_i}{\tau s + 1}$$

As $s \to 0$ then $sF(s)$ tends to $G\theta_i$ and so this is the final steady value. This is thus $30 \times 10^{-6} \times 100 = 300 \times 10^{-6}$ V.

Ramp input to a first-order system

Consider the behaviour of a first-order system when subject to a step input. For a first-order system we can take the relationship to be of the form given by equation [4], namely

$$G(s) = \frac{G}{\tau s + 1}$$

The Laplace transform of the output is thus

$$G(s) \times \text{Laplace transform of input}$$

$$\frac{G}{\tau s + 1} \times \text{Laplace transform of input}$$

The Laplace transform for a one unit slope ramp input at $t = 0$ is $1/s^2$. Hence, for such an input

$$\text{Laplace transform of output} = \frac{G}{(\tau s + 1)s^2}$$

$$= G \times \frac{(1/\tau)}{[s + (1/\tau)]s^2}$$

This transform is of the form

$$\frac{a}{s^2(s + a)}$$

which has the solution

$$t - \frac{(1 - e^{-at})}{a}$$

Thus the output θ_o for a unit slope ramp is given by

$$\theta_o = G[t - \tau(1 - e^{-t/\tau})] \qquad [11]$$

Figure 5.3 is a graph of this equation. It can be considered to be a graph of Gt minus a graph of $G\tau(1 - e^{-t/\tau})$. For a ramp input of slope A, i.e. $\theta_i = At$, then

$$\theta_o = GA[t - \tau(1 - e^{-t/\tau})] \qquad [12]$$

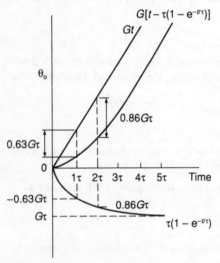

Fig. 5.3 $\theta_o = G[t - \tau(1 - e^{-t/\tau})]$ for unit ramp input

Example 4

A thermocouple has the transfer function linking its output in volts to its input θ_i in °C of

$$G(s) = \frac{30 \times 10^{-6}}{10s + 1}$$

When the thermocouple is subject to a steadily rising temperature input of 5°C/s, what will be the thermocouple output after 12 s and how much will it lag behind the output it would have indicated if it had responded instantly to the input?

Answer

For a ramp input to a first-order system, equation [12] gives

$$\theta_o = GA[t - \tau(1 - e^{-t/\tau})]$$

and, since the time constant τ for the system is 10 s (see Example 12) and G is 30×10^{-6} V/°C, then for a ramp of 5°C/s

$$\theta_o = 30 \times 10^{-6} \times 5[12 - 10(1 - e^{-12/10})] = 7.5 \times 10^{-4} \, \text{V}$$

As Fig. 5.3 indicates, the lag is the difference between the values of GAt and $GA[t - \tau(1 - e^{-t/\tau})]$. Since $GAt = 30 \times 10^{-6} \times 5 \times 12 = 18.0 \times 10^{-4}\,V$, the lag is $10.5 \times 10^{-4}\,V$.

Impulse input to a first-order system

Consider the behaviour of a first-order system when subject to an impulse input. For a first-order system we can take the relationship to be of the form given by equation [4], namely

$$G(s) = \frac{G}{\tau s + 1}$$

The Laplace transform of the output is thus

$$G(s) \times \text{Laplace transform of input}$$

$$\frac{G}{\tau s + 1} \times \text{Laplace transform of input}$$

The Laplace transform for a unit impulse at $t = 0$ is 1. Hence, for such an input

$$\text{Laplace transform of output} = \frac{G}{\tau s + 1} \times 1$$

$$= G \times \frac{(1/\tau)}{[s + (1/\tau)]}$$

This transform is of the form

$$\frac{1}{s + a}$$

and this is given by the function e^{-at}. Thus

$$\theta_o = G(1/\tau)\,e^{-t/\tau} \qquad [13]$$

Figure 5.4 is a graph of this equation. If the impulse has a size A, then

$$\theta_o = GA(1/\tau)\,e^{-t/\tau} \qquad [14]$$

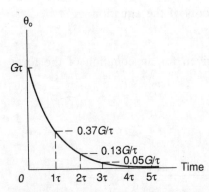

θ_o

$G\tau$

— $0.37G/\tau$

— $0.13G/\tau$

— $0.05G/\tau$

Time

$0 \quad 1\tau \quad 2\tau \quad 3\tau \quad 4\tau \quad 5\tau$

Fig. 5.4 $\theta_o = G(1/\tau)\,e^{-t/\tau}$ for unit impulse at $t = 0$

Example 5

A thermocouple has the transfer function linking its output in volts to its input θ_i in °C of

$$G(s) = \frac{30 \times 10^{-6}}{10s + 1}$$

What will be the output of the thermocouple 5 s after it was subject to a temperature impulse of 100 °C by suddenly and very briefly coming into contact with a hot object?

Answer

The thermocouple is a first-order system subject to an impulse of size

100 °C. Thus equation [14] can be used

$$\theta_o = GA(1/\tau)e^{-t/\tau}$$

and since $G = 30 \times 10^{-6}$ V and $\tau = 10$ s (see Example 3) then

$$\theta_o = 30 \times 10^{-6} \times 100(1/10)e^{-5/10} = 1.8 \times 10^{-4}$$ V

Step input to a second-order system   Consider the output from a second-order system when it is subject to a unit step input.

Laplace transform of output

$$= G(s) \times \text{Laplace transform of input}$$

Thus, using equation [8] to represent the general form of a second-order system in the s domain,

$$\theta_o(s) = \frac{b_0\omega_n^2}{s^2 + 2\zeta\omega_n s + \omega_n^2} \times \theta_i(s) \tag{15}$$

Since, for a unit step, $\theta_i(s) = 1/s$ then

$$\theta_o(s) = \frac{b_0\omega_n^2}{(s^2 + 2\zeta\omega_n s + \omega_n^2)s} \tag{16}$$

This can be rearranged as

$$\theta_o(s) = \frac{b_0\omega_n^2}{(s - m_1)(s - m_2)s} \tag{17}$$

where m_1 and m_2 are the roots of the equation

$$s^2 + 2\zeta\omega_n s + \omega_n^2 = 0$$

Thus, since the roots are given for an equation of the form $ax^2 + bx + c = 0$ by

$$x = \frac{-b \pm \sqrt{(b^2 - 4ac)}}{2a}$$

then

$$m = \frac{-2\zeta\omega_n \pm \sqrt{(4\zeta^2\omega_n^2 - 4\omega_n^2)}}{2}$$

$$m_1 = -\zeta\omega_n + \omega_n\sqrt{(\zeta^2 - 1)} \tag{18}$$

$$m_2 = -\zeta\omega_n - \omega_n\sqrt{(\zeta^2 - 1)} \tag{19}$$

The type of response that occurs, i.e. the inverse transformation, depends on the value of the damping factor ζ. When $\zeta > 1$ then $\sqrt{(\zeta^2 - 1)}$ is a real number and the system is said to be *overdamped*. This means both the roots are real.

Using partial fractions equation [21] can be rearranged as

$$\theta_o(s) = \frac{1}{s} + \frac{A}{s - m_1} + \frac{B}{s - m_2} \tag{20}$$

with

$$(s - m_1)(s - m_2) + As(s - m_2) + Bs(s - m_1) = b_0\omega^2 \quad [21]$$

Hence when $s = m_1$, then

$$Am_1(m_1 - m_2) = b_0\omega_n^2$$

$$A = \frac{b_0\omega_n^2}{m_1(m_1 - m_2)}$$

Substituting for the values of m_1 and m_2 from equations [18] and [19],

$$A = \frac{b_0\omega_n^2}{[-\zeta\omega_n + \omega_n\sqrt{(\zeta^2 - 1)}][2\,\omega_n\sqrt{(\zeta^2 - 1)}]}$$

$$A = \frac{b_0}{[-\zeta + \sqrt{(\zeta^2 - 1)}]2\sqrt{(\zeta^2 - 1)}}$$

Multiplying the top and bottom of this fraction by $[-\zeta - \sqrt{(\zeta^2 - 1)}]$ gives

$$A = \frac{b_0[-\zeta - \sqrt{(\zeta^2 - 1)}]}{[1]2\sqrt{(\zeta^2 - 1)}}$$

Hence

$$A = -\frac{b_0\zeta}{2\sqrt{(\zeta^2 - 1)}} - \frac{b_0}{2} \quad [22]$$

With equation [21] when $s = m_2$ then

$$Bm_2(m_2 - m_1) = b_0\omega_n^2$$

$$B = \frac{b_0\omega_n^2}{m_2(m_2 - m_1)}$$

and by a similar discussion to that above for A,

$$B = \frac{b_0\zeta}{2\sqrt{(\zeta^2 - 1)}} - \frac{b_0}{2} \quad [23]$$

The response of the system is the inverse transformation of equation [20]. The inverse transformation for $1/s$ is 1, for $A/(s - m_1)$ is $A\exp m_1 t$, and for $B/(s - m_2)$ is $B\exp m_2 t$, then

$$\theta_o = 1 + A\exp(m_1 t) + B\exp(m_2 t) \quad [24]$$

Substituting the values for A, B, m_1 and m_2 obtained above,

$$\theta_o = 1 + \left[-\frac{b_0\zeta}{2\sqrt{(\zeta^2 - 1)}} - \frac{b_0}{2}\right]$$

$$\exp\{[-\zeta\omega_n + \omega_n\sqrt{(\zeta^2 - 1)}]t\}$$

$$+ \left[\frac{b_0\zeta}{2\sqrt{(\zeta^2 - 1)}} - \frac{b_0}{2}\right]$$

$$\exp\{[-\zeta\omega_n - \omega_n\sqrt{(\zeta^2 - 1)}]t\} \quad [25]$$

When $\zeta = 1$ the system is said to be *critically damped*. For this condition $m_1 = m_2 = -\zeta\omega_n$. Equation [17] then becomes

$$\theta_o(s) = \frac{b_0\omega_n^2}{(s + \omega_n)^2 s}$$

The inverse transformation of this is (see Table 4.1)

$$\theta_o = b_0[1 - \exp(-\omega_n t) - \omega_n t \exp(-\omega_n t)] \qquad [26]$$

When $\zeta < 1$ the roots are complex and the system is said to be *underdamped*. When this happens the roots, equations [18] and [19] can be written as

$$m_1 = -\zeta\omega_n + \omega_n \sqrt{(\zeta^2 - 1)}$$

$$m_1 = -\zeta\omega_n + \omega_n \sqrt{[(-1)(1 - \zeta^2)]}$$

and thus writing j for $\sqrt{(-1)}$

$$m_1 = -\zeta\omega_n + j\omega_n \sqrt{(1 - \zeta^2)} \qquad [27]$$

Similarly since

$$m_2 = -\zeta\omega_n - \omega_n \sqrt{(\zeta^2 - 1)}$$

then

$$m_2 = -\zeta\omega_n - j\omega_n \sqrt{(1 - \zeta^2)} \qquad [28]$$

The inverse transform for equation [16] with this condition is (see Table 4.1)

$$\theta_o(s) = \frac{b_0\omega^2}{s(s^2 + 2\zeta\omega s + \omega^2)}$$

$$\theta_o = b_0\left[1 - \frac{1}{\sqrt{(1 - \zeta^2)}}\right.$$

$$\left. \exp(-\zeta\omega_n t)\sin[\omega_n \sqrt{(1 - \zeta^2)}t + \phi]\right] \qquad [29]$$

where $\cos\phi = \zeta$. This equation can be rearranged to give the form of equation quoted in Chapter 3. When $\zeta = 0$, i.e. there is no damping, then equation [28] gives

$$\theta_o = b_0\{1 - 1\,e^0 \sin[\omega_n \sqrt{(1)}t + 0]\}$$

$$\theta_o = b_0\{1 - \sin\omega_n t\}$$

The output thus oscillates with the undamped frequency ω_n.

Figure 3.12 shows graphs of the output responses at different values of damping factor, illustrating the effects of overdamping, critical damping and underdamping.

Example 6

A system has the following relationship, in the s domain, between its output θ_o and its input θ_i. What is the state of damping of the system when it is subject to a step input?

$$\frac{\theta_o(s)}{\theta_i(s)} = \frac{1}{s^2 + 8s + 16}$$

Answer

For a unit step input $\theta_i = 1/s$, hence

$$\theta_o = \frac{1}{s(s^2 + 8s + 16)}$$

This can be simplified to

$$\theta_o = \frac{1}{s(s + 4)(s + 4)}$$

The roots of the $s^2 + 8s + 16$ equation are thus $m_1 = m_2 = -4$. Both roots are real and the same. The system is thus critically damped.

Example 7

A second-order system is underdamped with a damping factor of 0.4 and a free angular frequency of 10 Hz. What is (a) the relationship between the output and the input in the s domain, (b) the relationship between the output and the input in the time domain when it is subject to a unit step input and (c) the percentage overshoot with such an input?

Answer

(a) In the s domain the second-order equation will be of the form given by equation [8]

$$G(s) = \frac{\theta_o(s)}{\theta_i(s)} = \frac{b_0\omega_n^2}{s^2 + 2\zeta\omega_n s + \omega_n^2}$$

where b_0 is a constant, ω_n the natural angular frequency and ζ the damping constant. Thus

$$\frac{\theta_o(s)}{\theta_i(s)} = \frac{100b_0}{s^2 + 8s + 100}$$

(b) When subject to a unit step input

$$\theta_o(s) = \frac{100b_0}{s(s^2 + 8s + 100)}$$

This has the general solution, as given by Table 4.1 and discussed above (equation [28]), of

$$\theta_o = b_0\left\{1 - \frac{1}{\sqrt{(1 - \zeta^2)}} \exp\left(-\zeta\omega_n t\right) \sin\left[\omega_n \sqrt{(1 - \zeta^2)}t + \phi\right]\right\}$$

where $\cos\phi = \zeta$. Thus, with $\omega_n = 10$ and $\zeta = 0.4$,

$$\theta_o = b_0\left\{1 - \frac{1}{0.84} e^{-4t} \sin\left(9.2t + 66.4°\right)\right\}$$

(c) As derived in Chapter 3, equation [33],

$$\text{Percentage overshoot} = \exp\left[\frac{-\zeta\pi}{\sqrt{(1-\zeta^2)}}\right] \times 100\%$$

and thus with $\zeta = 0.4$,

$$\text{Percentage overshoot} = \exp\left[\frac{-0.4\pi}{\sqrt{(1-0.4^2)}}\right] \times 100\%$$

$$= 25.4\%$$

Ramp input to a second-order system

Consider the output from a second-order system when it is subject to a unit ramp input.

Laplace transform of output

$= G(s) \times$ Laplace transform of input

Thus, using equation [8] to represent the general form of a second-order system in the s domain,

$$\theta_o(s) = \frac{b_0\omega_n^2}{s^2 + 2\zeta\omega_n s + \omega_n^2} \times \theta_i(s)$$

Since, for a unit ramp, $\theta_i(s) = 1/s^2$, then

$$\theta_o(s) = \frac{b_0\omega_n^2}{(s^2 + 2\zeta\omega_n s + \omega_n^2)s^2} \qquad [30]$$

$$\theta_o(s) = \frac{b_0\omega_n^2}{(s - m_1)(s - m_2)s^2} \qquad [31]$$

where m_1 and m_2 are the roots of the quadratic expression. Since for an equation of the form $ax^2 + bx + c$ the roots are given by

$$m = \frac{-b \pm \sqrt{(b^2 - 4ac)}}{2a}$$

then

$$m_1 = \frac{-2\zeta\omega_n + \sqrt{(4\zeta^2\omega_n^2 - 4\omega_n^2)}}{2} = -\zeta\omega_n + \omega_n\sqrt{(\zeta^2 - 1)}$$

and

$$m_2 = -\zeta\omega_n - \omega_n\sqrt{(\zeta^2 - 1)}$$

Equation [31] can be rearranged using partial fractions into the form

$$\theta_o(s) = b_0\left(\frac{A}{s^2} + \frac{B}{s} + \frac{C}{s - m_1} + \frac{D}{s - m_2}\right) \qquad [32]$$

Evaluation of these constants A, B, C and D and substitution of the values of m_1 and m_2 derived above gives

$$A = 1$$

$$B = -\frac{2\zeta}{\omega_n}$$

$$C = \frac{\zeta}{\omega_0} + \frac{2\zeta^2 - 1}{2\,\omega_n\sqrt{(\zeta^2 - 1)}}$$

$$D = \frac{\zeta}{\omega_n} - \frac{2\zeta^2 - 1}{2\,\omega_n\sqrt{(\zeta^2 - 1)}}$$

The inverse transformation of equation [32] gives

$$\theta_o = b_0[At + B + C\exp(m_1 t) + D\exp(m_2 t)]$$

$$\theta_o = b_0\left[t - \frac{2\zeta}{\omega_n} + C\exp(m_1 t) + D\exp(m_2 t)\right] \qquad [33]$$

The C and D terms in the equation give the transient response. The form of this response depends on whether ζ is greater than, equal to or less than 1 and so consequently the roots are real and unequal, real and equal, or complex and unequal. The form the transient response takes is thus the same as that occurring with the step input discussed earlier in this chapter. The A and B terms give the steady-state response. With no damping, i.e. $\zeta = 0$, then the steady-state response is just $b_0 t$ and indicates that the output keeps up with the steadily changing unit ramp input signal of t. However, when there is damping the steady-state response lags behind the input signal by $2\zeta/\omega_n$ (Fig. 5.5). This is referred to as the *steady-state error*. Figure 5.6 shows the types of steady-state plus transient responses that can occur with different degrees of damping.

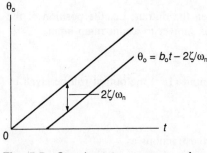

Fig. 5.5 Steady-state response of a second-order system to a unit ramp input

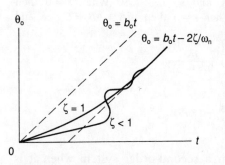

Fig. 5.6 Response of a second-order system to a unit ramp input

Example 8

The relationship between the input signal to a radiotelescope dish and the direction in which it points is given by the transfer function

$$G(s) = \frac{\omega_n^2}{s^2 + 2\zeta\omega_n + \omega_n^2}$$

with ζ having the value 0.4 and ω_n the value 10 Hz. What is (*a*) the steady-state error when the input signal to the telescope is a ramp signal, and (*b*) the 2% settling time?

Answer

(*a*) The steady-state error is the amount by which the steady-state position of the telescope lags behind the input ramp signal. This error has the value

$$\text{Steady-state error} = \frac{2\zeta}{\omega_n} = \frac{2 \times 0.4}{10} = 0.08\,\text{s}$$

(b) The settling time is the time taken for the response to settle down to within some fixed percentage of the steady-state value, in this case 2% (see Chapter 3 and equation [35]). This time is given by

$$t_s = \frac{4}{\zeta\omega_n} = \frac{4}{0.4 \times 10} = 1\,\text{s}$$

Example 9

A robot arm has a transfer function given by

$$G(s) = \frac{K}{(s + 3)^2}$$

Derive the relationship between the output, i.e. the position of the arm, and time when the arm is subject to a unit ramp input.

Answer

When subject to a unit ramp input ($1/s^2$) the output will be given by

$$\theta_o(s) = \frac{K}{s^2(s + 3)^2}$$

This can be rearranged in partial fractions as

$$\frac{A}{s^2} + \frac{B}{(s + 3)} + \frac{C}{(s + 3)^2}$$

Hence

$$A(s + 3)^2 + Bs^2(s + 3) + Cs^2 = K$$

When $s = -3$ then $9C = K$ and so $C = K/9$. When $s = 0$ then $9A = K$ and so $A = K/9$. When $s = 1$ then $16A + 4B + C = K$ and so $B = -2K/9$. Hence

$$\theta_o(s) = \frac{K}{9s^2} - \frac{2K}{9(s + 3)} + \frac{K}{9(s + 3)^2}$$

Using Table 4.1,

$$\theta_o = (K/9)t - (2K/9)\,e^{-3t} + (K/9)t\,e^{-3t}$$

Impulse input to a second-order system

Consider the output from a second-order system when it is subject to a unit impulse input at $t = 0$.

Laplace transform of output

$$= G(s) \times \text{Laplace transform of input}$$

Thus, using equation [8] to represent the general form of a second-order system in the s domain,

$$\theta_o(s) = \frac{b_0\omega_n^2}{s^2 + 2\zeta\omega_n s + \omega_n^2} \times \theta_i(s)$$

Since, for a unit impulse at $t = 0$, $\theta_i(s) = 1$ then

$$\theta_o(s) = \frac{b_0\omega_n^2}{(s^2 + 2\zeta\omega_n s + \omega_n^2)} \qquad [34]$$

This can be rearranged as

$$\theta_o(s) = \frac{b_0\omega_n^2}{(s - m_1)(s - m_2)} \qquad [35]$$

where m_1 and m_2 are the roots of the equation

$$s^2 + 2\zeta\omega_n s + \omega_n^2 = 0$$

Thus, since the roots are given for an equation of the form $ax^2 + bx + c = 0$ by

$$x = \frac{-b \pm \sqrt{(b^2 - 4ac)}}{2a}$$

then

$$m = \frac{-2\zeta\omega_n \pm \sqrt{(4\zeta^2\omega_n^2 - 4\omega_n^2)}}{2}$$

$$m_1 = -\zeta\omega_n + \omega_n\sqrt{(\zeta^2 - 1)}$$

$$m_2 = -\zeta\omega_n - \omega_n\sqrt{(\zeta^2 - 1)}$$

Using partial fractions equation [35] can be rearranged as

$$\theta_o(s) = \frac{A}{s - m_1} + \frac{B}{s - m_2} \qquad [36]$$

Hence

$$A(s - m_2) + B(s - m_1) = b_0\omega_n^2$$

and when $s = m_2$ then $B(m_2 - m_1) = b_0\omega_n^2$ and when $s = m_1$ then $A(m_1 - m_2) = b_0\omega_n^2$. Thus, substituting for the values of m_1 and m_2,

$$A = -B = \frac{b_0\omega_n}{2\sqrt{(\zeta^2 - 1)}}$$

The inverse transform of equation [36] gives

$$\theta_o = A\exp m_1 t + B\exp m_2 t$$

and thus, substituting for the values of A and B,

$$\theta_o = \frac{b_0\omega_n}{2\sqrt{(\zeta^2 - 1)}}[\exp(m_1 t) - \exp(m_2 t)] \qquad [37]$$

The form of response that occurs will depend on whether the roots m_1 and m_2 are real or complex, i.e. whether ζ is greater than or less than 1. When $\zeta > 1$ the roots are real and unequal and the result is just an increase in output which is followed by a slow decrease back to the zero value (Fig. 5.7). When $\zeta = 1$ the roots are real and equal and the system is critically

$\theta_o/b_o\omega_n$

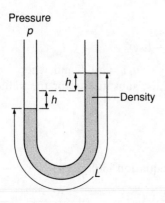

Fig. 5.7 Response of a second-order system to a unit impulse at $t = 0$

Pressure
p

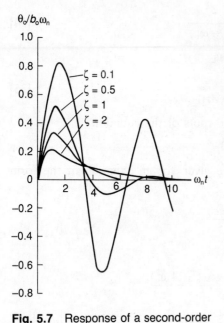

Fig. 5.8 Example 10

damped. This means that following the initial increase in output the response dies back to the zero value in the minimum time with no oscillations. When $\zeta < 1$ the roots are complex quantities and following the initial increase in output oscillations of steadily decreasing amplitude occur until eventually the output goes back to the zero value.

When ζ is less than 1 the roots are complex. When this is the case equation [37] is often written in a different form. This form could have been obtained by using the inverse transformation, given in Table 4.1, of

$$\frac{\omega_n^2}{s^2 + 2\zeta\omega_n s + \omega_n^2}$$

as

$$\frac{\omega_n}{\sqrt{(1 - \zeta^2)}} \exp(-\zeta\omega_n t) \sin[\omega_n \sqrt{(1 - \zeta^2)}t]$$

then, for equation [34],

$$\theta_o = \frac{b_o\omega_n}{\sqrt{(1 - \zeta^2)}} \exp(-\zeta\omega_n t) \sin[\omega_n \sqrt{(1 - \zeta^2)}t] \qquad [38]$$

Example 10

Figure 5.8 shows a simple second-order system of a liquid in a U-tube. Such a system is characteristic of systems involving liquids stored in two connected tanks and the sloshing of liquid between them that can occur. For the U-tube the transfer function relating a pressure input p to the liquid in one of the limbs to the output of a change in height h is given by

$$\frac{H(s)}{P(s)} = \frac{1}{\varrho L s^2 + Rs + 2\varrho g}$$

where ϱ is the liquid density, L the total length of the liquid column, R the hydraulic resistance and g the acceleration due to gravity. What is (a) the undamped frequency, (b) the damping ratio, and (c) the equation relating h to time when there is an impulse input?

Answer

(a) By comparison with equation [8]

$$G(s) = \frac{\theta_o(s)}{\theta_i(s)} = \frac{b_o\omega_n^2}{s^2 + 2\zeta\omega_n s + \omega_n^2}$$

and rearrangement of the U-tube transfer function equation

$$\frac{H(s)}{P(s)} = \frac{1}{\varrho L[s^2 + (R/\varrho L)s + (2g/L)]}$$

then $\omega_n = \sqrt{(2g/L)}$.

(b) The comparison also gives $2\zeta\omega_n = (R/\varrho L)$ and so

$$\zeta = \frac{R}{2\omega_n \varrho L} = \frac{R}{2\varrho L \sqrt{(2g/L)}} = \frac{R\sqrt{2}}{\varrho\sqrt{(Lg)}}$$

(c) The U-tube transfer function equation can thus be written as

$$\frac{H(s)}{P(s)} = \frac{(1/2\varrho g)\omega_n^2}{s^2 + 2\zeta\omega_n s + \omega_n^2}$$

Thus for an impulse input of pressure p then, since the problem states that sloshing of the liquid occurs, i.e. there are oscillations and so ζ is less than 1, the solution is of the form given in equation [38], i.e.

$$\theta_o = \frac{b_0\omega_n}{\sqrt{(1 - \zeta^2)}} \exp(-\zeta\omega_n t) \sin[\omega_n\sqrt{(1 - \zeta^2)}t]$$

Hence substituting the values of the U-tube,

$$h = (1/2\,\rho g)\sqrt{(2g/L)} \exp\left\{-\left[\frac{R\sqrt{2}}{\rho\sqrt{(Lg)}}\sqrt{(2g/L)}\right]t\right\}$$

$$\sin\left(\sqrt{(2g/L)}\sqrt{\left\{1 - \left[\frac{R\sqrt{2}}{\rho\sqrt{(Lg)}}\right]^2\right\}}\right)t$$

$$h = \frac{1}{\rho\sqrt{(2gL)}} \exp\left[\left(\frac{2R}{\rho L}\right)t\right] \sin\sqrt{\left(\frac{2g}{L} \quad \frac{4gR^2}{\rho^2 L^3}\right)}t$$

Problems

1 Write down the transfer function $G(s)$ for systems giving the following input-output relationships:

(a) RC circuit with input v output i

$$v = Ri + \frac{1}{C}\int i\,dt$$

(b) RLC circuit with input v output v_C

$$v = \frac{v_C}{R} + C\frac{dv_C}{dt} + \frac{1}{L}\int v_C dt$$

(c) Hydraulic system, a U-tube with input q output h

$$p = \varrho L\frac{d^2h}{dt^2} + RA\frac{dh}{dt} + 2h\varrho g$$

(d) Spring-dashpot-mass system with input F output x

$$\frac{1}{\omega_n^2}\frac{d^2x}{dt^2} + \frac{2\zeta}{\omega_n}\frac{dx}{dt} + x = \frac{F}{k}$$

2 What are the orders of the elements given the transfer functions obtained as answers in Problem 1 above?

3 Describe the behaviour of a first-order system when subject to (a) a step input, (b) a ramp input and (c) an impulse input.

4 A system has the following relationship between its output θ_o and its input θ_i in the s domain. What is the state of the damping in the system when it is subject to a step input?

$$\frac{\theta_o(s)}{\theta_i(s)} = \frac{1}{s^2 - 6s - 16}$$

5 A second-order system has a damping factor of 0.2, a free angular frequency of 5 Hz and a steady-state transfer function of 2. What is the relationship between the output and the input in the *s* domain for the system and the percentage overshoot when it is subject to a step input?

6 Describe the behaviour of a second-order system when subject to (*a*) a step input, (*b*) a ramp input and (*c*) an impulse input.

7 What is the response of the system giving the following transfer function when it is subject to a unit step input?

$$\frac{s}{(s + 3)^2}$$

8 A system has the following relationship between its output θ_o and its input θ_i in the *s* domain. What is the steady-state error when it is subject to a ramp input?

$$\frac{\theta_o}{\theta_i} = \frac{400}{s^2 + 20s + 400}$$

9 The side-to-side oscillations of a ship due to waves, i.e. the rolling motion, can be described by

$$\frac{\theta(s)}{H(s)} = \frac{\omega_n^2}{s^2 + 2\zeta\omega_n s + \omega_n^2}$$

where θ is the angular deflection from the vertical and *h* the height of the waves. With the free angular frequency ω_n as 2 Hz and the damping factor ζ as 0.1 how does the angular deflection vary with time for a sudden large wave, i.e. an impulse?

10 What is the response of the system giving the following transfer function when subject to a unit impulse?

$$G(s) = \frac{2}{(s + 3)(s + 4)}$$

6 Block diagram models

Introduction

Establishing the models for complex systems results in the linking together of a number of subsystems or elements, each of which has its own transfer function. Block diagrams can be used to represent each of these subsystems and, the grouped and linked array, the system as a whole. This chapter is concerned with such groupings and how the overall response of the system be determined from a knowledge of the transfer functions of the individual, constituent blocks.

The block diagram

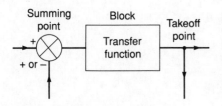

Fig. 6.1   Block diagram items

Figure 6.1 shows the way the various items in block diagrams are represented. *Arrows* are used to represent the directions of signal flow. Where signals are functions of time they are represented by lower-case letters followed by (t), as for example $i(t)$, though often the (t) will be omitted where it is obvious that the signals are functions of time. Where signals are in the s domain they are represented by capital-letters followed by (s), as for example $I(s)$. A *summing point* is where signals are algebraically added together. If one of the signals entering such a point is marked as being positive and the other is marked as negative then the sum is the difference between the two signals. If both the signals are marked as positive then the sum is the addition of the two signals. In Chapter 1 such a summing point when used to compare the required value with a feedback signal indicating the actual value was referred to as the *comparator*, the feedback signal being subtracted from the required value to give the error signal. Where a signal is taken off from some point in a signal path the *take-off point* is represented in the same way as when, in an electrical circuit, there is a junction between conductors which enables a current to be taken off, i.e. the junction is represented by two lines meeting and the junction marked with a '·'. The *block* is usually drawn with its transfer function written inside it.

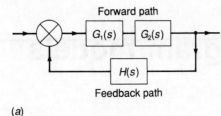

Forward path

Feedback path

(a)

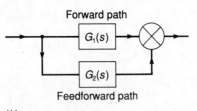

Forward path

Feedforward path

(b)

Fig. 6.2 Signal paths

Blocks in series

The term *forward path* is used for those elements through which the signal passes when moving in the direction input-to-output for the system as a whole (Fig. 6.2(a)). The transfer functions for elements in this forward path are usually designated by G or $G(s)$. The term *feedback path* is used for those elements through which a signal passes when being fed back from the output towards the input (Fig. 6.2(a)). The transfer functions for elements in this feedback path are usually designated by H or $H(s)$. The term *feedforward path* is used for those elements which are in parallel with the forward path and through which signals move in the same direction, i.e. input-to-output (Fig. 6.2(b)).

If a system consists of a number of elements in series, i.e. as the two elements in forward path in Fig. 6.2(a), then the transfer function $G(s)$ of the system is

$$G(s) = \frac{\theta_o(s)}{\theta_i(s)}$$

But for the first element we have

$$G_1(s) = \frac{\theta_{o1}(s)}{\theta_i(s)}$$

and for element 2

$$G_2(s) = \frac{\theta_{o2}(s)}{\theta_{o1}(s)}$$

and for element 3

$$G_3(s) = \frac{\theta_o(s)}{\theta_{o2}(s)}$$

But

$$\frac{\theta_o(s)}{\theta_i(s)} = \frac{\theta_{o1}(s)}{\theta_i(s)} \times \frac{\theta_{o2}(s)}{\theta_{o1}(s)} \times \frac{\theta_o(s)}{\theta_{o2}(s)}$$

Thus

$$G(s) = G_1(s) \times G_2(s) \times G_3(s) \qquad [1]$$

Thus a number of blocks in series, with transfer functions $G_1(s)$, $G_2(s)$, $G_3(s)$, etc., can be replaced by a single block with a transfer function $G(s)$. See Chapter 1 for the derivation of the above when the transfer function referred only to the steady state.

Thus for two elements $G_1(s)$ and $G_2(s)$ in series, if $\theta_i(s)$ is the input to the arrangement and $\theta_o(s)$ the output

$$\theta_o(s) = G(s)\theta_i(s) = G_1(s)G_2(s)\theta_i(s) \qquad [2]$$

It is assumed that when the individual blocks are linked together that there is no interaction between the blocks which results in changes in the transfer function of the individual blocks. Only if there is no such interaction can the transfer functions of the blocks when isolated be used to derive the overall transfer function when the blocks are combined. Thus if the individual blocks are electrical circuits there can be problems in combining them because the circuits interact and load each other.

Example 1

An open-loop system consists of two elements in series, the elements having the transfer functions indicated in Fig. 6.3. What is the transfer function of the system as a whole?

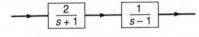

Fig. 6.3 Example 1

Answer

For blocks in series equation [1] gives

$$G(s) = G_1(s) \times G_2(s) \times G_3(s) \times \text{etc.}$$

Hence

$$G(s) = \frac{2}{s+1} \times \frac{1}{s-1} = \frac{2}{s^2-1}$$

Blocks with feedback loop

Figure 6.4(a) shows a simple closed-loop system with negative feedback. With negative feedback the reference and the feedback signals are subtracted at the summing point, with positive feedback they are added. If $\theta_i(s)$ is the reference value, i.e. the input, and $\theta_o(s)$ the actual value, i.e. the output, of the system then the transfer function of the entire control system is

$$\text{Transfer function} = \frac{\theta_i(s)}{\theta_o(s)}$$

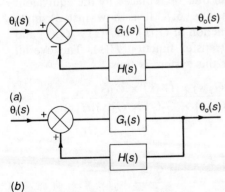

(a)

(b)

Fig. 6.4 (a) Negative feedback, (b) positive feedback

Each subsystem within the overall system has its own transfer function. For the feedback path, transfer function $H(s)$, then with its input of $\theta_o(s)$ there will be an output of $H(s)\theta_o(s)$ into the forward path. Thus if the forward path system has a transfer function $G_1(s)$ then with its input of $\theta_i(s) - H(s)\theta_o(s)$ and output of $\theta_o(s)$,

$$G_1(s) = \frac{\theta_o(s)}{\theta_i(s) - H(s)\theta_o(s)}$$

Thus, rearranging this gives

$$\theta_o(s)[1 + G_1(s)H(s)] = \theta_i(s)G_1(s) \qquad [3]$$

Hence the overall transfer function of the closed-loop control system is

$$G(s) = \frac{G_1(s)}{1 + G_1(s)H(s)} \qquad [4]$$

If the feedback is positive (Fig. 6.4(b)), then

$$\theta_o(s) = G_1(s)[\theta_i(s) + G_1(s)H(s)\theta_o(s)] \qquad [5]$$

$$G(s) = \frac{G_1(s)}{1 - G_1(s)H(s)} \qquad [6]$$

Example 2

What is the overall transfer function for the system shown in Fig. 6.5?

Answer

The feedback is negative, hence using equation [4]

$$G(s) = \frac{G_1(s)}{1 + G_1(s)H(s)}$$

$$G(s) = \frac{2/(s + 1)}{1 + 5s[2/(s + 1)]} = \frac{2}{(s + 1) + 10s} = \frac{2}{1 + 11s}$$

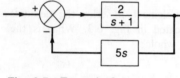

Fig. 6.5   Example 2

Blocks in series and with a feedback loop

Consider a closed-loop system consisting of three components in series for the forward path and with a feedback loop, as shown in Fig. 6.6. The forward-path transfer function is thus

Forward-path transfer function $= G_1(s) \times G_2(s) \times G_3(s)$

The closed-loop system can thus be replaced by the equivalent simpler system shown in Fig. 6.7. It is now just a single element with a transfer function of $G_1(s) \times G_2(s) \times G_3(s)$ and a feedback loop with a transfer function $H(s)$. The overall transfer function $G(s)$ for this system is thus

$$G(s) = \frac{\theta_o(s)}{\theta_i(s)} = \frac{G_1(s) \times G_2(s) \times G_3(s)}{1 + [G_1(s) \times G_2(s) \times G_3(s)]H(s)}$$

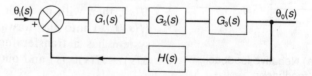

Fig. 6.6   The transfer function with a multi-element closed-loop system

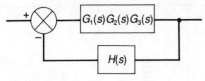

Fig. 6.7 The equivalent system for Fig. 6.6

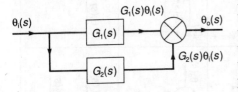

Fig. 6.8 Example 3

Blocks in parallel

$\theta_i(s)$

$G_1(s)\theta_i(s)$

$G_1(s)$

$\theta_o(s)$

$G_2(s)$

$G_2(s)\theta_i(s)$

Fig. 6.9 A feedforward loop

Example 3

A control system (Fig. 6.8) has a forward path of two elements, transfer functions K and $1/(s + 1)$. If the feedback path has a transfer function of s, what is the transfer function of the closed-loop system.

Answer

For the forward-path elements the overall transfer function is, using equation [1],

$$\text{Forward-path transfer function } (s) = K \times \frac{1}{s + 1}$$

Since there is negative feedback, then the overall transfer function $G(s)$ is, using equation [4],

$$G(s) = \frac{G_1(s)}{1 + G_1(s)H(s)}$$

$$G(s) = \frac{K/(s + 1)}{1 + [sK/(s + 1)]} = \frac{K}{1 + s(1 + K)}$$

Figure 6.9 shows part of a control system which has a feedforward loop. For such a system the signal input to each of the elements is $\theta_i(s)$. Thus the output from the element with the transfer function $G_1(s)$ is $G_1(s)\theta_i(s)$ and the output from the element with the transfer function $G_2(s)$ is $G_2(s)\theta_i(s)$. In the figure the two signals are shown as adding at the summing point. Hence the output $\theta_o(s)$ is

$$\theta_o(s) = G_1(s)\theta_i(s) + G_2(s)\theta_i(s) = [G_1(s) + G_2(s)]\theta_i(s) \quad [7]$$

Hence the overall transfer function $G(s)$ is

$$G(s) = G_1(s) + G_2(s) \quad [8]$$

If the signals had been subtracted at the summing point then we would have had

$$\theta_o(s) = [G_1(s) - G_2(s)]\theta_i(s) \quad [9]$$

and

$$G(s) = G_1(s) - G_2(s) \quad [10]$$

Example 4

A feedforward loop of the form shown in Fig. 6.9 has transfer functions of $G_1(s) = 1/(s + 1)$ and $G_2(s) = 5$. What is the overall transfer function if the signal from the feedforward loop adds to the forward path signal?

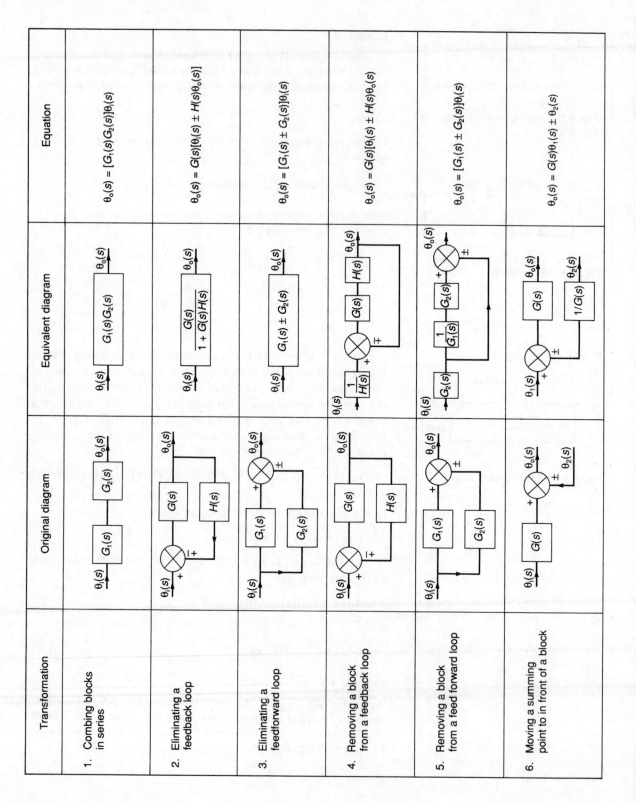

Transformation	Original diagram	Equivalent diagram	Equation
1. Combing blocks in series			$\theta_o(s) = [G_1(s)G_2(s)]\theta_i(s)$
2. Eliminating a feedback loop			$\theta_o(s) = G(s)[\theta_i(s) \pm H(s)\theta_o(s)]$
3. Eliminating a feedforward loop			$\theta_o(s) = [G_1(s) \pm G_2(s)]\theta_i(s)$
4. Removing a block from a feedback loop			$\theta_o(s) = G(s)[\theta_i(s) \pm H(s)\theta_o(s)]$
5. Removing a block from a feed forward loop			$\theta_o(s) = [G_1(s) \pm G_2(s)]\theta_i(s)$
6. Moving a summing point to in front of a block			$\theta_o(s) = G(s)\theta_1(s) \pm \theta_2(s)$

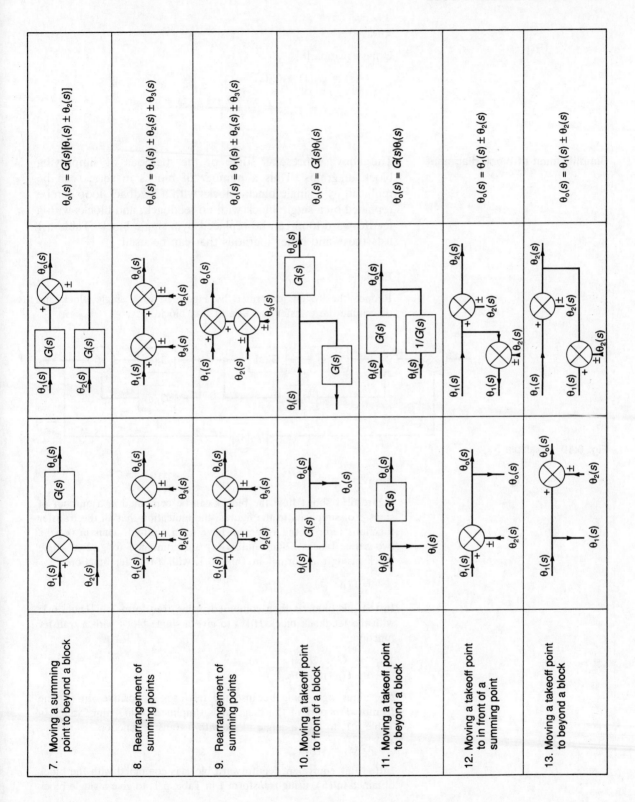

7.	Moving a summing point to beyond a block		$\theta_o(s) = G(s)[\theta_1(s) \pm \theta_2(s)]$
8.	Rearrangement of summing points		$\theta_o(s) = \theta_1(s) \pm \theta_2(s) \pm \theta_3(s)$
9.	Rearrangement of summing points		$\theta_o(s) = \theta_1(s) \pm \theta_2(s) \pm \theta_3(s)$
10.	Moving a takeoff point to front of a block		$\theta_o(s) = G(s)\theta_1(s)$
11.	Moving a takeoff point to beyond a block		$\theta_o(s) = G(s)\theta_1(s)$
12.	Moving a takeoff point to in front of a summing point		$\theta_o(s) = \theta_1(s) \pm \theta_2(s)$
13.	Moving a takeoff point to beyond a block		$\theta_o(s) = \theta_1(s) \pm \theta_2(s)$

Answer

Using equation [8]

$$G(s) = G_1(s) + G_2(s)$$

$$G(s) = 5 + \frac{1}{s+1} = \frac{5s + 5 + 1}{s+1} = \frac{5s + 6}{s+1}$$

Simplification of block diagrams

The above represents some of the methods of simplifying block diagrams. Thus a number of blocks in series can be replaced by a single block, blocks with a feedback loop can be replaced by a single block with no feedback, and blocks with a feedforward loop can be replaced by a single block. Table 6.1 lists these and other methods that can be used.

Example 5

Reduce the system described by Fig. 6.10 to a single block and determine the transfer function of that block.

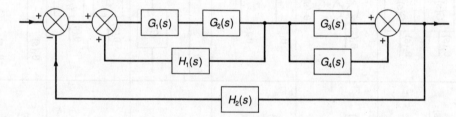

Fig. 6.10 Example 5

Answer

Figure 6.11 shows how the blocks can be combined in a number of steps. For simplicity in the figures, the indication that all the transfer functions refer to the s-domain have been omitted. Thus in (a) the two series blocks $G_1(s)$ and $G_2(s)$ are combined to give a single block, using transform 1 in Table 6.1, with a transfer function

$$G_1(s)G_2(s)$$

In (b) this block is then combined, using transform 2 in Table 6.1, with the feedback block $H_1(s)$ to give a single block with a transfer function

$$\frac{G_1(s)G_2(s)}{1 - G_1(s)G_2(s)H_1(s)}$$

the minus sign being because the feedback is positive. In (c) the feedforward part of the diagram is simplified, using transform 3 in Table 6.1, to give a single block with a transfer function

$$G_1(s) + G_2(s)$$

In (d) this equivalent feedforward block is combined with the block obtained in (b), using transform 1 in Table 6.1, to give a single block

Diagram blocks   Equivalent

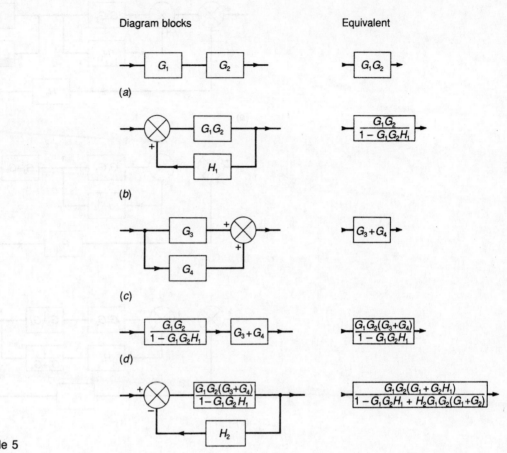

Fig. 6.11   Example 5

with a transfer function

$$\frac{G_1(s)G_2(s)[G_1(s) + G_2(s)]}{1 - G_1(s)G_2(s)H_1(s)}$$

Finally in (e) this block is combined with its feedback loop, using transform 2 in Table 6.1, to give a single block with a transfer function

$$\frac{G_1(s)G_2(s)[G_1(s) + G_2(s)]/[1 - G_1(s)G_2(s)H_1(s)]}{1 + H_2(s)\{G_1(s)G_2(s)[G_1(s) + G_2(s)]/[1 - G_1(s)G_2(s)H_1(s)]\}}$$

This simplifies to

$$\frac{G_1(s)G_2(s)[G_1(s) + G_2(s)]}{1 - G_1(s)G_2(s)H_1(s) + H_2(s)G_1(s)G_2(s)[G_1(s) + G_2(s)]}$$

Example 6

Reorganize the block diagram of the system described in Fig. 6.10 so that feedback box $H_1(s)$ is isolated and effects of changes in its transform function can be more readily examined.

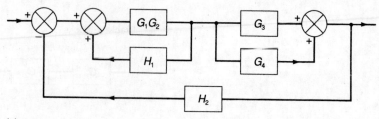

(a)

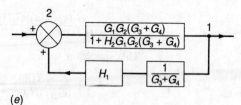

(b)

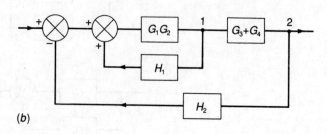

(c)

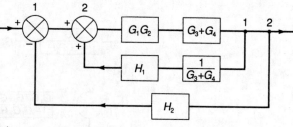

(d)

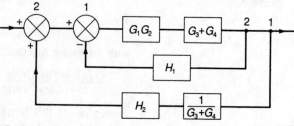

(e)

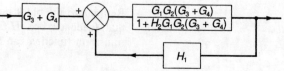

(f)

Fig. 6.12 Example 6

Answer

Figure 6.12 shows the steps in the procedure that can be adopted to isolate block $H_1(s)$. For simplicity the indication that all the transfer functions are in the s-domain has been omitted from the figures. In (a) the two blocks in series have been combined using transform 1 from Table 6.1. In (b) the feedforward loop has been eliminated using transform 3 from Table 6.1. In (c) the take-off point 1 has been moved to beyond block $[G_3(s) + G_4(s)]$, using transform 11 from Table 6.1. In (d) the summing points 1 and 2 have been rearranged, using transform 9 from Table 6.1. In (e) the inner feedback loop has been eliminated using transform 2 from Table 6.1. In (f) transform 4 has been used to remove a block from the feedback loop and so give the required answer.

Example 7

Transform the system shown in Fig. 6.13 into one with unity feedback. Unity feedback is when the feedback path $H(s)$ is transformed to one with a feedback transfer function of 1.

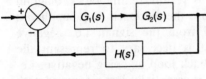

Fig. 6.13 Example 7

Answer

Using transformation 4 in Table 6.1 results in the system shown in Fig. 6.14.

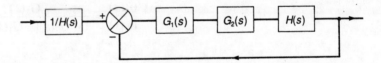

Fig. 6.14 Example 7

Multiple inputs

Often in control systems there are more than one input to the system. Thus there can be the input signal indicating the required value of the controlled variable and also an input or inputs due to disturbances which affect the system. The procedure that can be adopted to obtain the relationship between the inputs and the output for such systems is:

1   Set all but one of the inputs to zero.
2   Transform the resulting block diagram to one with just a forward path and a feedback path.
3   Then determine the output signal due to this one non-zero input.
4   Repeat the above steps 1, 2 and 3 for each of the inputs in turn.
5   The total output of the system is the algebraic sum of the outputs due to each of the inputs.

Figure 6.15 shows a basic control system with the reference signal input $\theta_i(s)$ and a disturbance input $\theta_d(s)$. Applying the above procedures gives, with $\theta_d(s)$ equated to zero and after some simplification, the block diagram shown in Fig. 6.16(a). For such a system the relationship between the input $\theta_i(s)$ and the output $\theta_o(s)$ is, using transformation 2 in Table 6.1,

$$\frac{\theta_o(s)}{\theta_i(s)} = \frac{G_1(s)G_2(s)}{1 + G_1(s)G_2(s)H(s)} \quad [11]$$

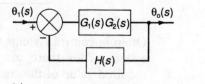

Fig. 6.15 A control system with multiple inputs

Now putting $\theta_i(s)$ equal to 0 results in the block diagram shown in Fig. 6.16(b). Because the feedback loop supplied a signal which was subtracted from the signal, i.e. negative feedback was being used, it is necessary to represent the transfer function of the feedback loop as being negative, i.e. $-H(s)$. The resulting system is essentially just a forward path with transfer function $G_2(s)$ and positive feedback with a transfer function of $-G_1(s)H(s)$. Thus, using transformation 2 in Table 6.1,

$$\frac{\theta_o(s)}{\theta_d(s)} = \frac{G_2(s)}{1 - G_2(s)[-G_1(s)H(s)]} = \frac{G_2(s)}{1 + G_1(s)G_2(s)H(s)} \quad [12]$$

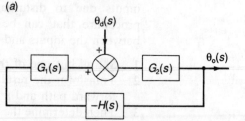

(a)

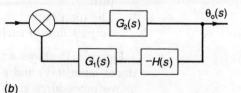

Fig. 6.16 (a) $\theta_d(s) = 0$, (b) $\theta_i(s) = 0$ (b)

Thus the total output of the system when subject to both inputs is the sum of that given in equations [11] and [12], i.e.

$$\theta_o(s) = \frac{G_1(s)G_2(s)\theta_i(s)}{1 + G_1(s)G_2(s)H(s)} + \frac{G_2(s)\theta_d(s)}{1 + G_1(s)G_2(s)H(s)} \quad [13]$$

See Chapter 1 for a derivation of the above equation in terms of the signals passing through the system.

Example 8

Derive an equation describing the relationship between the inputs $\theta_i(s)$, $\theta_{d1}(s)$ and $\theta_{d2}(s)$ to the system described in Fig. 6.17 and the output $\theta_o(s)$.

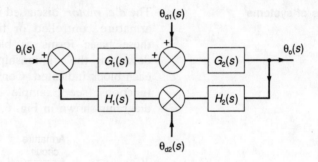

Fig. 6.17 Example 8

Answer

With $\theta_{d1}(s)$ and $\theta_{d2}(s)$ both zero, transformation 2 Table 6.1 gives

$$\frac{\theta_o(s)}{\theta_i(s)} = \frac{G_1(s)G_2(s)}{1 + G_1(s)G_2(s)H_1(s)H_2(s)}$$

With $\theta_i(s)$ and $\theta_{d2}(s)$ both zero the arrangement is as in Fig. 6.18(a) and so

$$\frac{\theta_o(s)}{\theta_{d1}(s)} = \frac{G_2(s)}{1 + G_1(s)G_2(s)H_1(s)H_2(s)}$$

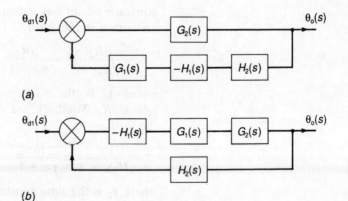

(a)

(b)

Fig. 6.18 Example 8

With $\theta_i(s)$ and $\theta_{d1}(s)$ both zero the arrangement is as in Fig. 6.18(b) and so

$$\frac{\theta_o(s)}{\theta_{d2}(s)} = -\frac{G_1(s)G_2(s)H_1(s)}{1 + G_1(s)G_2(s)H_1(s)H_2(s)}$$

Hence the total output of the system is

$$\theta_o(s) = \frac{G_1(s)G_2(s)\theta_i(s)}{1 + G_1(s)G_2(s)H_1(s)H_2(s)} + \frac{G_2(s)\theta_{d1}(s)}{1 + G_1(s)G_2(s)H_1(s)H_2(s)} - \frac{G_1(s)G_2(s)H_1(s)\theta_{d2}(s)}{1 + G_1(s)G_2(s)H_1(s)H_2(s)}$$

Examples of systems

The *d.c. motor*, discussed in Chapter 2, can have two forms, as armature controlled or field controlled. Figure 2.39 shows these basic forms as block diagrams with the equations describing the relationships between the input and output to each block indicated. Converting these equations into transfer functions (see Example 1(f) in Chapter 5) gives the block diagrams shown in Fig. 6.19.

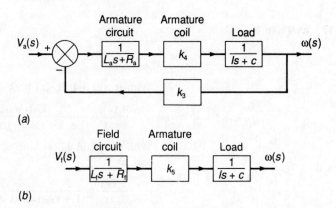

(a)

(b)

Fig. 6.19 d.c. motor, (a) armature controlled, (b) field controlled

For the *armature controlled motor* (Fig. 6.19(a)), the armature circuit has a first-order transfer function which can be written in the form

$$\frac{1/R_a}{(L_a/R_a) + 1} = \frac{1/R_a}{\tau_1 s + 1}$$

where τ_1 is the time constant for the armature circuit. $\tau_1 = L_a/R_a$. Similarly, we can write the transfer function for the load as

$$\frac{1/c}{(I/c)s + 1} = \frac{1/c}{\tau_2 s + 1}$$

where τ_2 is the time constant for the load, $\tau_2 = I/c$.
The forward-path transfer function for the system is thus

$$\frac{1/R_a}{\tau_1 s + 1} \times k_4 \times \frac{1/c}{\tau_2 s + 1}$$

The transfer function of the system with its feedback path $\omega(s)/V_a(s)$ is thus, using transformation 2 from Table 6.1,

$$G(s) = \frac{(1/R_a)k_4(1/c)/(\tau_1 s + 1)(\tau_2 s + 1)}{1 + k_3(1/R_a)k_4(1/c)/(\tau_1 s + 1)(\tau_2 s + 1)}$$

$$= \frac{(1/R_a)k_4(1/c)}{(\tau_1 s + 1)(\tau_2 s + 1) + k_3(1/R_a)k_4(1/c)} \qquad [14]$$

Equation [14] can be rearranged to give

$$G(s) = \frac{(1/R_a)k_4(1/c)}{\tau_1\tau_2 s^2 + (\tau_2 + \tau_2)s + 1 + k_3(1/R_a)k_4(1/c)}$$

$$= \frac{[(1/R_a)k_4(1/c)]/(\tau_1\tau_2)}{s^2 + [(\tau_1 + \tau_2)/(\tau_1\tau_2)]s + [k_3(1/R_a)k_4(1/c) + 1]/(\tau_1\tau_2)}$$

This is a second-order equation and can be written in the form

$$G(s) = \frac{K}{s^2 + 2\zeta\omega_n s + \omega_n^2}$$

where ω_n is the natural angular frequency and ζ the damping ratio. The behaviour of the system when subject to, for example, a step or ramp input is thus as described in Chapter 5 for second-order systems.

For the *field-controlled motor* (Fig. 6.19(b)), the field circuit is a first-order system and the transfer function can be written as

$$\frac{1}{L_f s + R_f} = \frac{1/R_f}{\tau_1 s + 1}$$

where the time constant $\tau_1 = L_f/R_f$. The load system is also a first-order system and its transfer function can be written as

$$\frac{1/c}{(I/c)s + 1} = \frac{1/c}{\tau_2 s + 1}$$

where τ_2 is the time constant for the load, $\tau_2 = I/c$.

The field-controlled motor is an open loop system. Hence the overall transfer function $\omega(s)/v_f(s)$ is

$$G(s) = \frac{1/R_f}{\tau_1 s + 1} \times k_5 \times \frac{1/c}{\tau_2 s + 1}$$

$$G(s) = \frac{(1/R_f)k_5(1/c)}{(\tau_1 s + 1)(\tau_2 s + 1)} \qquad [15]$$

This is a second-order system.

With d.c. motors the most likely disturbance that can occur is a torque T_d to the load. Thus to take account of this the block diagrams for the armature-controlled and field-controlled motors are modified to be as in Fig. 6.20. The effect of the disturbance on the armature controlled motor output is to modify the output indicated by equation [14] to

$$\omega(s) = \frac{(1/R_a)k_4(1/c)}{(\tau_1 s + 1)(\tau_2 s + 1) + k_3(1/R_a)k_4(1/c)}V_a(s)$$

$$+ \frac{(1/c)(\tau_1 s + 1)}{(\tau_1 s + 1)(\tau_2 s + 1) + (\tau_2 s + 1)k_3 k_4(1/R_a)}T_d(s)$$

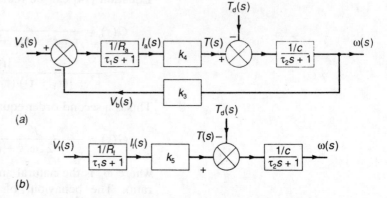

(a)

Fig. 6.20 d.c. motor with load disturbance, (a) armature controlled, (b) field controlled

(b)

A d.c. motor might be used as part of a *position-control system*, i.e. to rotate the load to some particular angle. Since the output from the d.c. motors above has been considered to be the angular velocity ω we need to consider adding a block which converts angular velocity to angular displacement. Since angular velocity is rate of angular displacement, i.e. ω = dθ/dt, then

$$\int d\theta = \int \omega dt$$

Hence in the s domain

$$\theta_o(s) = \frac{1}{s}\omega(s)$$

where $\theta_o(s)$ is the s domain output. Thus the block to be added has the transfer function 1/s, and hence for the armature-controlled motor is as shown in Fig. 6.21. The

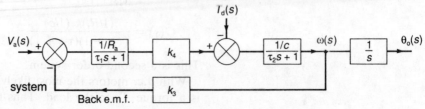

Fig. 6.21 Position-control system using a d.c. motor

feedback loop in this case is just the back e.m.f. and since this is a measure of the angular velocity the take-off point for the feedback is prior to the 1/s block where the output from the load is angular velocity rather than angular displacement.

The control performance of the position-control system can be improved by including a second feedback path, this being one which is a measure of the angular position θ_o of the load. Figure 6.22 shows the system. The measurement system used could be a rotary potentiometer or a rotary variable-differential transformer (RVDT) which give a voltage proportional to the angular position. This feedback signal v_m is summed with a reference voltage signal v_r to give the error signal. The reference voltage signal is obtained from the angular setting of a rotary potentiometer (see Chapter 2 for a discussion of the potentiometer). The error signal is amplified and then rectified before passing to the d.c. motor system. Such a control system could be used to control the movement of robot arms, position control with machine tools, etc.

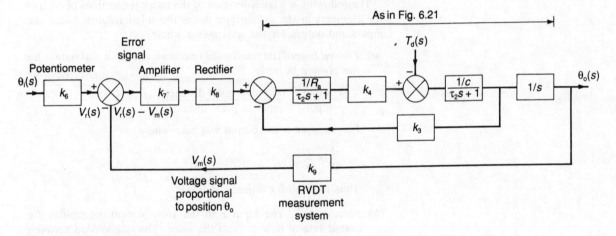

Fig. 6.22 Position-control system with position feedback

Example 9

Derive the relationship between the inputs of the set value for liquid height h_i and disturbance to the liquid level of d to the liquid level system described by Fig. 6.23 and the output of the liquid height h_o. Hence determine how the output will vary with time when there is a step input to set value, i.e. it is suddenly set to a new value.

Movement of the float results in a beam rotating about a pivot and so, via the actuator, adjusting the opening of the valve. The set point can be adjusted by raising or lowering the point of connection of the float to the beam. The error signal is the difference between the position of the float when at the required level and when it is at any other level. The controller is the pivoted beam with an input at the float end of the error signal and an output at the point where the actuator is connected.

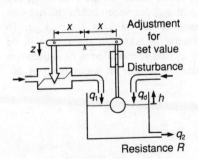

Fig. 6.23 Example 9

Answer

The system can be considered to have a block diagram like that shown in Fig. 6.24. The error signal is the input to one end of the pivoted beam and the output is the movement of the other end of the pivoted beam and hence the movement of the valve stem. This movement is the input to the valve actuator and results in an output of a flow rate. This, summed with the disturbance, is the input to the plant, i.e. the liquid in the container. The feedback is the movement of the float which directly translates into a height signal, hence the unity feedback path.

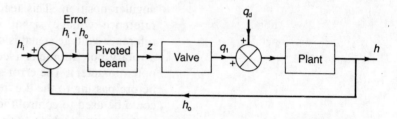

Fig. 6.24 Example 9

The following is a consideration of the transfer functions of each of the elements in the system and hence the relationship between the inputs and output for the system as a whole.

(a) *Pivoted beam*. The relationship between the input and output for the pivoted beam is

$$\frac{\text{output}}{\text{input}} = \frac{\text{output to pivot distance}}{\text{input to pivot distance}}$$

For an input h the output will be z where

$$\frac{z}{h} = \frac{x}{y}$$

Thus the transfer function is (x/y).

(b) *Valve system*. The input z to the valve system determines the output rate of flow q_1 from the valve. The relationship between input and output can be linearized, see Chapter 2, to give

$$q_1 = cz$$

(c) *The plant*. The level is controlled by control of the rate at which water enters the tank. If the water enters at the rate of q_{in} per second and leaves at the rate of q_{out} per second then the net rate at which the water increases in the tank is $(q_{in} - q_{out})$. In a time δt the change in volume of water in the tank will be $(q_{in} - q_{out})\delta t$. If the tank has a cross-sectional area of A then this change in volume will produce a change in water level of δh, where

$$(q_{in} - q_{out})\delta t = A\,\delta h$$

$$q_{in} - q_{out} = A\frac{\delta h}{\delta t}$$

Hence

$$q_{in} - q_{out} = A \frac{dh}{dt}$$

The outflow from the tank will be affected by the height of the water above the exit pipe, i.e. the pressure head. Since the pressure due to a height of water h is proportional to h we can write

$$q_{out} = \frac{h}{R}$$

where R represents the hydraulic resistance of the exit pipe. Thus

$$q_{in} - \frac{h}{R} = A \frac{dh}{dt}$$

$$RA \frac{dh}{dt} + h = Rq_{in}$$

The transfer function is thus

$$G(s) = \frac{H(s)}{Q_{in}(s)} = \frac{R}{RAs + 1} = \frac{R}{\tau s + 1}$$

where $\tau = RA$ and is the time constant.

(d) *The system.* With $q_d = 0$ the forward path has a relationship between output and input of

$$\frac{x}{y} \times c \times \frac{R}{\tau s + 1}$$

Since the feedback path has unity transfer function, then

$$H(s) = \frac{(x/y)c[R/(\tau s + 1)]}{1 + (x/y)c[R/(\tau s + 1)]} h_i(s)$$

With $q_i(s) = 0$ the forward path has a transfer function of $R/(\tau s + 1)$ and the feedback path one of $(x/y)c$. Hence

$$H(s) = \frac{R/(\tau s + 1)}{1 + (x/y)c[R/(\tau s + 1)]} Q_d(s)$$

Hence for the system with both inputs

$$H(s) = \frac{(x/y)c[R/(\tau s + 1)]}{1 + (x/y)c[R/(\tau s + 1)]} H_i(s)$$

$$+ \frac{R/(\tau s + 1)}{1 + (x/y)c[R/(\tau s + 1)]} Q_d(s)$$

(e) *Step input.* Consider a step input to the system, with no disturbance occurring. For this situation the relationship between the input and output is

$$H(s) = \frac{(x/y)c[R/(\tau s + 1)]}{1 + (x/y)c[R/(\tau s + 1)]} H_i(s)$$

$$H(s) = \frac{(x/y)cR}{\tau s + 1 + (x/y)cR} H_i(s)$$

This equation can be simplified if we define two constants a and K as

$$a = \frac{1 + (x/y)cR}{\tau}$$

$$K = \frac{(x/y)cR}{1 + (x/y)cR}$$

Hence

$$H(s) = \frac{Ka}{s + a} h_i(s)$$

A step input for $h_i(s)$ has the Laplace transform $1/s$, hence

$$H(s) = \frac{Ka}{s(s + a)}$$

The equation which would give this transform (see Chapter 4) is

$$h = K(1 - e^{-at})$$

This is an exponential growth to reach a steady state value of K.

Problems

1 What are the overall transfer functions of the systems described by the block diagrams in Fig. 6.25?

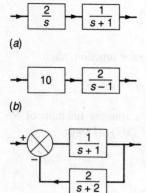

(a)

(b)

(c)

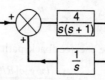

(d)

(e)

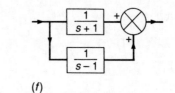

(f)

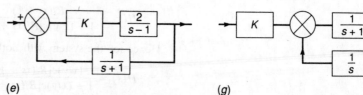

(g)

Fig. 6.25 Problem 1

2 Convert the block diagrams in Fig. 6.26 to the equivalent systems with unity feedback.

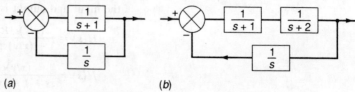

(a)

(b)

Fig. 6.26 Problem 2

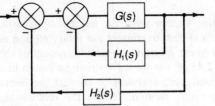

3 Verify the transformation given for removing a block from a feedback loop, transformation 4 Table 6.1.

4 Simplify Fig. 6.27 so that the forward path block $G(s)$ is isolated.

5 Derive an equation relating the inputs $\theta_i(s)$, $\theta_{d1}(s)$ and $\theta_{d2}(s)$ and the output for the system described by Fig. 6.28.

Fig. 6.27 Problem 4

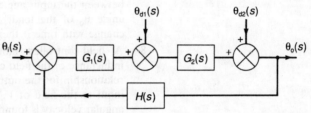

Fig. 6.28 Problem 5

6 Figure 6.29 shows a speed control system that might be used to control the speed of rotation of the wheels of a train locomotive. A diesel fuelled engine is used to drive an electric generator which in turn drives a motor and hence the wheels of the locomotive. The position of the throttle lever determines the reference voltage supplied to the differential amplifier. There it is compared with the feedback signal and an amplified error voltage is used to provide the current in the field coil of the generator. The generator output is controlled by the field current since the rotor is rotated at constant angular velocity. The output from the generator supplied the current in the armature circuit of a motor, the output from which is a torque which rotates the shaft on which the train wheels are located. Represent the system by means of a block diagram and derive an equation describing the relationship between the angle of the throttle lever θ and the angular speed output ω_o.

Fig. 6.29 Problem 6

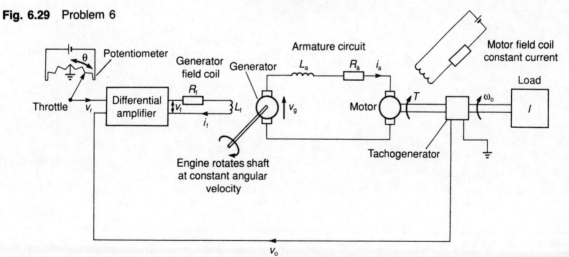

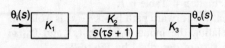

Fig. 6.30 Problem 7

7 A power-assisted steering system for a car has the basic form shown in Fig. 6.30, i.e. a system to convert the rotary motion of the steering wheel to a linear motion, a hydraulic system of the form described in Fig. 2.43 of a spool valve driving a piston in a cylinder and hence a load, and a system to convert the linear motion of the load to rotary motion. Derive the relationship between the input angle θ_i to the steering wheel and the output angle θ_o of the load. Hence state how the output angle will change with time if there is a unit ramp input.

8 A field-controlled motor is used to rotate a load. If the inductance of the field coils can be considered negligible derive a relationship for the output speed ω_o when there is a step voltage input to the field of V. If the step is 50 V then the steady-state angular velocity is found to be 2 rad/s. The load reaches 1 rad/s at 0.35 s after the step voltage was applied. Using this data derive the equation for the graph describing how the output speed varies with time.

7 Steady-state error

Introduction

When a command input is applied to a control system it is generally hoped that after any transient effects have died away the system output will settle down to the commanded value. The error between this value and the input command is called the *steady-state error*. It is a measure of the accuracy of a control system to track a command input and is the error after all transient responses to the input have decayed (see Chapter 5 for a preliminary discussion of this error in relation to the dynamic response of systems). The steady-state error for a system depends on the system concerned and the form taken for the system input. In order to carry out an analysis of the steady-state errors of systems it is useful to classify systems into *system types*. This type number indicates for each type of input the steady-state error that will occur. This chapter is about this classification and the determination of steady-state errors for systems.

Steady-state error

The error with any system is the difference between the required output signal, i.e. the reference input signal which specifies what is required, and the actual output signal that occurs. For an open-loop control system (Fig. 7.1) when there is an input $\theta_i(s)$ and an output $\theta_o(s)$ then the error $E(s)$ is

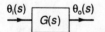

Fig. 7.1 Open-loop control system

$$E(s) = \theta_i(s) - \theta_o(s)$$

Since the transfer function for the $G(s)$ is $\theta_o(s)/\theta_i(s)$ then

$$E(s) = \theta_i - G(s)\theta_i(s) = [1 - G(s)]\theta_i(s) \qquad [1]$$

The error thus depends not only on the system, as determined by its transfer function, but also the form of the input $\theta_i(s)$.

For a closed-loop system, consider a simplification of one with unity feedback (Fig. 7.2). When there is a reference input $\theta_i(s)$ and an actual output $\theta_o(s)$ then the fed back signal is $\theta_o(s)$ and so the error $E(s)$ is

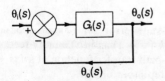

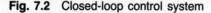

Fig. 7.2 Closed-loop control system

$$E(s) = \theta_i(s) - \theta_o(s)$$

If $G(s)$ is the forward-path transfer function then, since there is unity feedback,

$$\frac{\theta_o(s)}{\theta_i(s)} = \frac{G(s)}{1 + G(s)}$$

Hence

$$E(s) = \theta_i(s) - \frac{G(s)\theta_i(s)}{1 + G(s)} = \frac{1}{1 + G(s)}\theta_i(s) \qquad [2]$$

The error thus depends on the system, as specified by its transfer function, and the form of the input $\theta_i(s)$.

If the closed-loop system has a feedback loop with a transfer function $H(s)$, as in Fig. 7.3(a), then the system can be converted to one with unit feedback by the process outlined in Fig. 7.3(b). The result is an equivalent unity feedback system of the form indicated in Fig. 7.39(c). The forward-path transfer function is then given by

$$\frac{G(s)}{1 + G(s)[H(s) - 1]} \qquad [3]$$

Simplifying the system by converting it to unity feedback enables equation [2] to be used for the error.

In order to calculate the steady-state error e_{ss} the final-value theorem (see Chapter 4) is used. The steady-state error is the value of the error, which is a function of time t, when all transients have had time to decay, and thus the value as t tends to infinity. According to the final-value theorem this condition is given by

$$e_{ss} = \lim_{t \to \infty} e(t) = \lim_{s \to 0} sE(s) \qquad [4]$$

Thus for the open-loop system, using equation [1],

$$e_{ss} = \lim_{s \to 0} \{s[1 - G(s)]\theta_i(s)\} \qquad [5]$$

and for the closed-loop system, using equation [2],

$$e_{ss} = \lim_{s \to 0} \left[s\frac{1}{1 + G(s)}\theta_i(s) \right] \qquad [6]$$

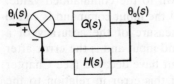

$\theta_i(s)$ $G(s)$ $\theta_o(s)$

$H(s)$

(a)

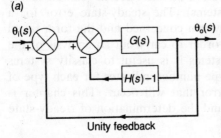

$\theta_i(s)$ $G(s)$ $\theta_o(s)$

$H(s) - 1$

Unity feedback

(b)

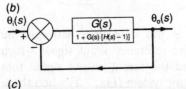

$\theta_i(s)$ $\dfrac{G(s)}{1 + G(s)\,[H(s) - 1]}$ $\theta_o(s)$

(c)

Fig. 7.3 (a) Closed-loop control system, (b) converting to unity feedback, (c) equivalent system with unity feedback

Example 1

Calculate the size of the steady-state error for (a) an open-loop system with a transfer function of $k/(\tau s + 1)$ and (b) a unity-feedback closed-loop system with a forward-path transfer function of $k/(\tau s + 1)$ when both are subject to a unit step input of $1/s$.

Answer

(*a*) Using equation [5]

$$e_{ss} = \lim_{s \to 0} \{s[1 - G(s)]\theta_i(s)\}$$

$$= \lim_{s \to 0} \left[s\left(1 - \frac{k}{\tau s + 1}\right)\frac{1}{s} \right]$$

$$= 1 - k$$

(*b*) Using equation [6]

$$e_{ss} = \lim_{s \to 0} \left[s\frac{1}{1 + G(s)}\theta_i(s) \right]$$

$$= \lim_{s \to 0} \left[s\frac{1}{1 + k/(\tau s + 1)}\frac{1}{s} \right]$$

$$= \frac{1}{1 + k}$$

System classification

The steady-state error for a system depends on the value of

$$\lim_{s \to 0} sE(s)$$

and the value of $E(s)$ depends on the forward-path transfer function of a closed-loop system when there is unity feedback. In discussions about the classification of systems it is important to realize that in all closed-loop cases the system is assumed to be in the form which gives unity feedback. Systems are classified according to the value of the forward-path transfer function when there is unity feedback, this often being called the *open-loop transfer function* of the closed-loop system. For a system with a forward transfer function $G(s)$ and feedback loop with transfer function $H(s)$, the open-loop transfer function $G_o(s)$ is given by equation [3] as

$$G_o(s) = \frac{G(s)}{1 + G(s)[H(s) - 1]} \tag{7}$$

The open-loop transfer function of systems can be represented in general by an equation of the form

$$\frac{K(s^m + a_{m-1}s^{m-1} + a_{m-2}s^{m-2} + \ldots a_1s + a_0)}{s^q(s^n + b_{n-1}s^{n-1} + b_{n-2}s^{n-2} + \ldots b_1s + b_0)} \tag{8}$$

where K is a constant and m and n are integers and neither a_0 or b_0 are zero. q is an integer, the value of which is called the *type* or *class* of the system. Thus if $q = 0$ then the system is said to be type 0, if $q = 1$ then type 1, if $q = 2$ then type 2.

The type number is thus the number of $1/s$ factors in the open-loop transfer function. Since $1/s$ is integration (see

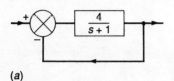

(a)

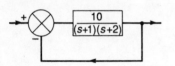

(b)

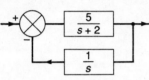

(c)

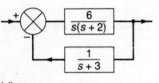

(d)

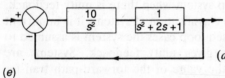

(e)

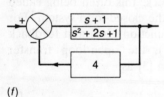

(f)

Fig. 7.4 Example 2

Chapter 4), the type number is the number of *integrators* in the open-loop transfer function.

Example 2

What are the type numbers for the systems shown in Fig. 7.4?

Answer

The open-loop transfer functions can be calculated using equation [7].

(a) This system has an open-loop transfer function of

$$\frac{4}{(s + 1)}$$

Since there is no independent s term in the denominator the type is 0.

(b) This system has an open-loop transfer function of

$$\frac{10}{(s + 1)(s + 2)}$$

Since there is no independent s term in the denominator the type is 0.

(c) This system has an open-loop transfer function of

$$\frac{5/(s + 2)}{1 + [5/(s + 2)][(1/s) - 1]} = \frac{5s}{(s^2 - 3s + 5)}$$

Since there is no independent s term in the denominator the type is 0.

(d) This system has an open-loop transfer function of

$$\frac{6/[s(s + 2)]}{1 + \{6/[s(s + 2)]\}[1/(s + 3) - 1]} = \frac{6(s + 3)}{s^3 + 5s^2 + 9s - 12}$$

Since there is no independent s term in the denominator the type is 0.

(e) This system has an open-loop transfer function of

$$\frac{10}{s^2(s^2 + 2s + 1)}$$

Since there is an independent s^2 term in the denominator the type is 2.

(f) This system has an open-loop transfer function of

$$\frac{(s + 1)/(s^2 + 2s + 1)}{1 + [(s + 1)/(s^2 + 2s + 1)](4 - 1)} = \frac{s + 1}{s^2 + 5s + 4}$$

Since there is no independent s term in the denominator the type is 0.

Steady-state error for a step input   The steady-state error e_{ss} for a closed-loop system is given by equation [6] as

$$e_{ss} = \lim_{s \to 0} \left[s \frac{1}{1 + G_o(s)} \theta_i(s) \right]$$

where $G_o(s)$ is the open-loop transfer function. A unit step input has $\theta_i(s) = 1/s$. Thus for such an input

$$e_{ss} = \lim_{s \to 0} \left[s \frac{1}{1 + G_o(s)} \frac{1}{s} \right]$$

$$= \lim_{s \to 0} \left[\frac{1}{1 + G_o(s)} \right] \qquad [9]$$

The open-loop transfer function is given by equation [8] as

$$\frac{K(s^m + a_{m-1}s^{m-1} + a_{m-2}s^{m-2} + \ldots a_1 s + a_0)}{s^q(s^n + b_{n-1}s^{n-1} + b_{n-2}s^{n-2} + \ldots b_1 s + b_0)}$$

and thus as s approaches zero the open-loop transfer function for a type 0 system will become Ka_0/b_0, i.e. a constant, and for all other types infinity. It is more usual to represent the value that the open-loop transfer function approaches as $s \to 0$ as a constant K_p, where K_p is called the *positional error constant* and has no units.

$$K_p = \lim_{s \to 0} G_o(s) \qquad [10]$$

Thus, in terms of the above equation for the open-loop transfer function

$$K_p = Ka_0/b_0 \qquad [11]$$

for a type 0 system and infinity for all other types.

The consequence of this is that the steady-state error for a type 0 system will be $1/[1 + (Ka_0/b_0)]$ or

$$e_{ss} = \frac{1}{1 + K_p} \qquad [12]$$

and for all other types zero. Figure 7.5 shows the type of response that might occur with a type 0 system. After the transients, whatever their form, have died away there is a steady-state error of $1/(1 + K_p)$.

The above represents the situation when there is a unit step input. If the input was a step of size A then the steady-state error with a type 0 system would be $A/(1 + K_p)$.

With a type 0 system the size of the steady-state error with a unit step input depends on the value of K_p: the bigger its value the smaller the error. But K_p is directly proportional to K (equation [11]). K is the factor by which signals are multiplied in passing through the forward path of the system. For example, the system might be like that shown in Fig. 7.6. Thus by increasing this amplification factor or gain the steady-state error can be reduced.

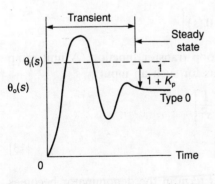

Fig. 7.5 Steady-state error with step input

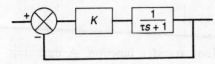

Fig. 7.6 A type 0 system

Example 3

What are the steady-state errors when a unit step input is applied to systems giving the following open-loop transfer functions?

(a) $\dfrac{4}{s+1}$

(b) $\dfrac{10}{(s+1)(s+2)}$

(c) $\dfrac{5}{s(s^2-3s+5)}$

(d) $\dfrac{6(s+3)}{(s+2)(s-6)}$

(e) $\dfrac{10}{s^2(s^2+2s+1)}$

Answer

(a) This is a type 0 system. As $s \to 0$ then $G_o(s)$ tends to 4. Thus $K_p = 4$ and so the steady-state error is $1/(1+4) = 0.2$ units.

(b) This is a type 0 system. As $s \to 0$ then $G_o(s)$ tends to 5. Thus $K_p = 5$ and so the steady-state error is $1/(1+5) = 0.17$ units.

(c) This is a type 1 system and so the steady-state error is zero.

(d) This is a type 0 system. As $s \to 0$ then $G_o(s)$ tends to $-3/2$. Thus $K_p = -3/2$ and so the steady-state error is $1/(1-1.5) = -2.0$ units.

(e) This is a type 2 system and so the steady-state error is zero.

Steady-state error for a ramp input

The steady-state error e_{ss} for a closed-loop system is given by equation [6] as

$$e_{ss} = \lim_{s \to 0} \left[s \frac{1}{1+G_o(s)} \theta_i(s) \right]$$

where $G_o(s)$ is the open-loop transfer function. A unit ramp input has $\theta_i(s) = 1/s^2$. Thus for such an input

$$e_{ss} = \lim_{s \to 0} \left[s \frac{1}{1+G_o(s)} \frac{1}{s^2} \right]$$

$$= \lim_{s \to 0} \left[\frac{1}{s + sG_o(s)} \right] \tag{13}$$

As s tends to zero so the s term in the denominator becomes zero. The factor thus determining the size of the error is thus the value of $sG_o(s)$ as $s \to 0$, i.e. equation [13] becomes

$$e_{ss} = \frac{1}{\lim_{s \to 0} sG_o(s)} \tag{14}$$

$$e_{ss} = \frac{1}{K_v} \tag{15}$$

where K_v is a constant, known as the *velocity error constant*. It has the units of second^{-1}.

$$K_v = \lim_{s \to 0} sG_o(s) \tag{16}$$

The open-loop transfer function G_o is given by equation [8] as

$$\frac{K(s^m + a_{m-1}s^{m-1} + a_{m-2}s^{m-2} + \ldots a_1s + a_0)}{s^q(s^n + b_{n-1}s^{n-1} + b_{n-2}s^{n-2} + \ldots b_1s + b_0)}$$

Thus the value of $sG_o(s)$ is

$$\frac{sK(s^m + a_{m-1}s^{m-1} + a_{m-2}s^{m-2} + \ldots a_1s + a_0)}{s^q(s^n + b_{n-1}s^{n-1} + b_{n-2}s^{n-2} + \ldots b_1s + b_0)}$$

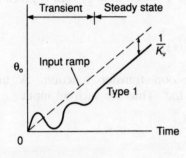

Fig. 7.7 Steady-state error with ramp input

For a type 0 system $q = 0$, hence $sK/s^q = sK$. Thus as s approaches zero $sG_o(s)$ for a type 0 system will become zero and so K_v zero. Thus the value of the steady-state error will be 1/0 or infinity. For a type 1 system $q = 1$, hence $sK/s^q = K$. Thus as s approaches zero $sG_o(s)$ will become Ka_0/b_0, i.e. this is the value of K_v. Thus the value of the steady-state error will be $1/K_v$ or $1/(Ka_0/b_0)$. Figure 7.7 shows the type of response that might occur with a type 1 system. After the transients, whatever their form, have died away there is a steady-state error of $1/K_v$. For a type 2 system $q = 2$, hence $sK/s^q = K/s$. Thus as s approaches zero $sG_o(s)$ becomes infinity and hence the steady-state error becomes zero.

The above represents the situation when there is a unit ramp input. If the input was a ramp with a rate of change with time of a constant A then the steady-state error with the type 1 system would be A/K_v.

Example 4

What are the steady-state errors when a unit ramp input is applied to systems having the following open-loop transfer functions? They are the systems considered in Example 3.

(a) $\dfrac{4}{s + 1}$

(b) $\dfrac{10}{(s + 1)(s + 2)}$

(c) $\dfrac{5}{s(s^2 - 3s + 5)}$

(d) $\dfrac{6(s + 3)}{(s + 2)(s - 6)}$

(e) $\dfrac{10}{s^2(s^2 + 2s + 1)}$

Answer

(a) This is a type 0 system. As $s \to 0$ then $sG_o(s)$ tends to 0. Thus $K_v = 0$ and so the steady-state error is infinite.

(b) This is a type 0 system. As $s \to 0$ then $sG_o(s)$ tends to 0. Thus $K_v = 0$ and so the steady-state error is infinite.

(c) This is a type 1 system. As $s \to 0$ then $sG_o(s)$ tends to the value $5/5 = 1\,s^{-1}$ and so the steady-state error is 1 unit.

(d) This is a type 0 system. As $s \to 0$ then $sG_o(s)$ tends to 0. Thus $K_v = 0$ and so the steady-state error is infinite.

(e) This is a type 2 system. As $s \to 0$ then $sG_o(s)$ tends to infinity and so the steady-state error is zero.

Steady-state error for a parabolic input

The steady-state error e_{ss} for a closed-loop system is given by equation [6] as

$$e_{ss} = \lim_{s \to 0} \left[s \frac{1}{1 + G_o(s)} \theta_i(s) \right]$$

where $G_o(s)$ is the open-loop transfer function. A unit parabolic input has $\theta_i(s) = 1/s^3$. Thus for such an input

$$e_{ss} = \lim_{s \to 0} \left[s \frac{1}{1 + G_o(s)} \frac{1}{s^3} \right]$$

$$= \lim_{s \to 0} \left[\frac{1}{s^2 + s^2 G_o(s)} \right] \tag{17}$$

As s tends to zero so the s^2 term in the denominator becomes zero. The factor thus determining the size of the error is thus the value of $s^2 G_o(s)$ as $s \to 0$, i.e. equation [17] becomes

$$e_{ss} = \frac{1}{\lim_{s \to 0} s^2 G_o(s)} \tag{18}$$

$$e_{ss} = \frac{1}{K_a} \tag{19}$$

where K_a is a constant, known as the *acceleration error constant*, and has the units of second^{-2}.

$$K_a = \lim_{s \to 0} s^2 G_o(s) \tag{20}$$

The open-loop transfer function G_o is given by equation [8] as

$$\frac{K(s^m + a_{m-1}s^{m-1} + a_{m-2}s^{m-2} + \ldots a_1 s + a_0)}{s^q(s^n + b_{n-1}s^{n-1} + b_{n-2}s^{n-2} + \ldots b_1 s + b_0)}$$

Thus the value of $s^2 G_o(s)$ is

$$\frac{s^2 K(s^m + a_{m-1}s^{m-1} + a_{m-2}s^{m-2} + \ldots a_1 s + a_0)}{s^q(s^n + b_{n-1}s^{n-1} + b_{n-2}s^{n-2} + \ldots b_1 s + b_0)}$$

For a type 0 system $q = 0$, hence $s^2 K/s^q = s^2 K$. Thus as s approaches zero $s^2 G_o(s)$ for a type 0 system will become zero and so K_a zero. Thus the value of the steady-state error will be 1/0 or infinity. For a type 1 system $q = 1$, hence $s^2 K/s^q = sK$. Thus as s approaches zero $s^2 G_o(s)$ will become zero and so K_a zero. Thus the value of the steady-state error will be 1/0 or infinity. For a type 2 system $q = 2$, hence $s^2 K/s^q = K$. Thus as s approaches zero $s^2 G_o(s)$ becomes Ka_0/b_0, i.e. this is the value of K_a. Thus the value of the steady-state error will be $1/K_a$ or $1/(Ka_0/b_0)$. Figure 7.8 shows the type of response that might occur with a type 2 system. After the transients, whatever their form, have died away there is a steady-state error of $1/K_v$. For systems with higher type numbers as s approaches zero $s^2 G_o(s)$ becomes infinity and hence the steady-state error zero.

The above represents the situation when there is a unit parabolic input. If the input was parabolic of the form A/s^3, where A is a constant, then the steady-state error with the type 2 system would be A/K_a.

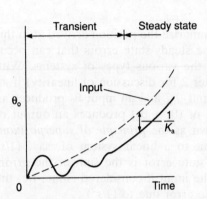

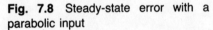

Fig. 7.8 Steady-state error with a parabolic input

Example 5

What are the steady-state errors when a unit parabolic input is applied to systems having the following open-loop transfer functions? They are the systems considered in Examples 3 and 4.

(a) $\dfrac{4}{s + 1}$

(b) $\dfrac{10}{(s + 1)(s + 2)}$

(c) $\dfrac{5}{s(s^2 - 3s + 5)}$

(d) $\dfrac{6(s + 3)}{(s + 2)(s - 6)}$

(e) $\dfrac{10}{s^2(s^2 + 2s + 1)}$

Answer

(a) This is a type 0 system. As $s \to 0$ then $s^2 G_o(s)$ tends to 0. Thus $K_a = 0$ and so the steady-state error is infinite.

(b) This is a type 0 system. As $s \to 0$ then $s^2 G_o(s)$ tends to 0. Thus $K_a = 0$ and so the steady-state error is infinite.

(c) This is a type 1 system. As $s \to 0$ then $s^2 G_o(s)$ tends to 0. Thus $K_a = 0$ and so the steady-state error is infinite.

(d) This is a type 0 system. As $s \to 0$ then $s^2 G_o(s)$ tends to 0. Thus

$K_a = 0$ and so the steady-state error is infinite.

(e) This is a type 2 system. As $s \to 0$ then $s^2 G_o(s)$ tends to $10 \mathrm{s}^{-2}$. Thus $K_a = 10 \mathrm{s}^{-2}$ and so the steady-state error is 1/10 unit.

(f) This is a type 0 system. As $s \to 0$ then $s^2 G_o(s)$ tends to 0. Thus $K_a = 0$ and so the steady-state error is infinite.

Steady-state errors for different inputs

Table 7.1 and Fig. 7.9 summarize the position arrived at in this chapter with regard to the steady-state errors that can occur with different inputs for the various types of systems. With linear systems (see Chapter 2 for discussion of linearity) if an input θ_1 produces an output θ_{o1} and an input θ_2 produces an output θ_{o2} then an input of $(\theta_1 + \theta_2)$ produces an output of $(\theta_{o1} + \theta_{o2})$. This is known as the *principle of superposition*. Thus if we have an input to a linear system of, say, $(1/s)$ + $(1/s^2)$ then the steady-state error is the sum of the errors due to each segment of the input if considered alone, i.e. the error due to $(1/s)$ plus the error due to $(1/s^2)$.

Table 7.1 Steady-state errors

| System type | Steady-state errors due to input of unit | | | |
	Step $1/s$	Ramp $1/s^2$	Parabola $1/s^3$	$1/s^4$
0	$1/(1 + K_p)$	∞	∞	∞
1	0	$1/K_v$	∞	∞
2	0	0	$1/K_a$	∞
3	0	0	0	$1/K_4$
4	0	0	0	0

Example 6

A motor control system is required for a computer disk drive. It is required to operate with zero steady-state error when there is a ramp input signal. What type number is required for the system?

Answer

For zero steady-state error with a ramp input signal, Table 7.1 indicates that the system must be type 2 or higher.

Example 7

A robot arm has an open-loop transfer function for its angular position of

$$G_o(s) = \frac{100}{s(s + 5)(s + 2)}$$

What will be the steady-state error when the input is as indicated in Fig. 7.10?

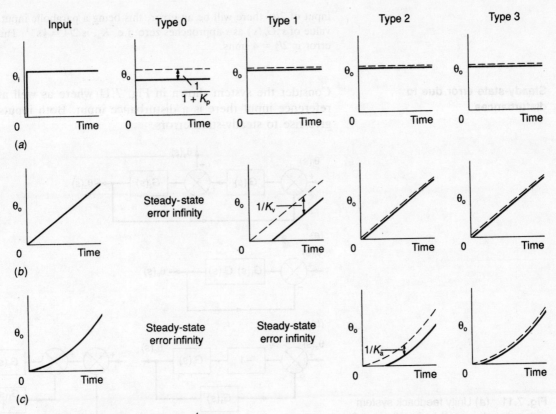

Input   Type 0   Type 1   Type 2   Type 3

Fig. 7.9 Steady-state errors: (a) step input, (b) ramp input and (c) parabolic input

Answer

This is a type 1 system and the input is a ramp signal of 10 degrees per second. As s tends to zero so $sG_o(s)$ tends to $100/10 = 10\,s^{-1}$. Thus $K_v = 10\,s^{-1}$ and so the steady-state error is, for unit ramp input, $1/K_v$ and for a ramp of rate A is A/K_v, i.e. $10/10 = 1°$. Thus the output is always lagging behind the input by $1°$.

Example 8

Determine the steady-state error that occurs with a linear system having the open-loop transfer function of

$$G_o(s) = \frac{2(s + 1)}{s^2(s + 4)}$$

when subject to an input of

$$\theta_i(s) = \frac{1}{s} + \frac{2}{s^2} + \frac{2}{s^3}$$

Fig. 7.10 Example 7

Answer

The system is type 2. Because the system is linear the principle of superposition can be applied and so the total steady-state error will be the sum of the errors due to each segment of the input signal. Thus for the input of $1/s$, a unit step input, there will be zero error. For the input of $2/s^2$ there will be no error, this being a ramp input. For the input of $2/s^3$ there will be an error, this being a parabolic input. The value of $s^2 G_o(s)$ as s approaches zero, i.e. K_a, is $2/4 = \frac{1}{2}s^{-1}$. Thus the error is $2/\frac{1}{2} = 4$ units.

Steady-state error due to disturbances

Consider the system shown in Fig. 7.11 where as well as the reference input there is a disturbance input. Both inputs can give rise to steady-state errors.

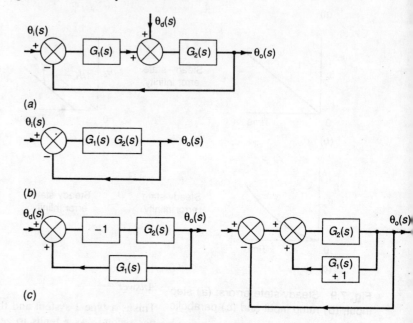

Fig. 7.11 (a) Unity feedback system with a disturbance, (b) when $\theta_d(s) = 0$, (c) when $\theta_i(s) = 0$

Following the procedure outlined in Chapter 6 the open-loop transfer function is determined for when $\theta_d(s) = 0$ but $\theta_i(s)$ is not zero and the steady-state error determined, then for when $\theta_i(s) = 0$ but $\theta_d(s)$ is not zero. The steady-state errors when both inputs are not zero is then sum of the two separately derived errors. Thus for $\theta_d(s) = 0$ we have

$$G_o(s) = \frac{G_1(s)G_2(s)}{1 + G_1(s)G_2(s)}$$

The error is the difference between the reference input and the output of the system

$$E(s) = \theta_i(s) - \theta_o(s)$$

and so, since $G_o(s) = \theta_o(s)/\theta_i(s)$,

$$E(s) = \theta_i(s) - \frac{G_1(s)G_2(s)}{1 + G_1(s)G_2(s)}\theta_i(s)$$

$$E(s) = \frac{1}{1 + G_1(s)G_2(s)}\theta_i(s)$$

Hence the steady-state error is

$$e_{ss} = \lim_{s \to 0} \left[s \frac{1}{1 + G_1(s)G_2(s)} \theta_i(s) \right] \qquad [21]$$

When $\theta_i(s) = 0$ then the system is a forward path of $G_2(s)$ and a feedback path of $G_1(s)$. This can be converted into a unity feedback system by the method outlined in Fig. 7.11, and so the open-loop transfer function is

$$G_o(s) = \frac{G_2(s)}{1 + G_2(s)(-1)[G_1(s) + 1]}$$

The error is thus, since $\theta_i(s) = 0$,

$$E(s) = \theta_i(s) - \theta_o(s) = -\theta_o(s)$$

$$E(s) = -\frac{G_2(s)}{1 + G_2(s)[G_1(s) + 1]} \theta_d(s)$$

Hence the steady-state error is

$$e_{ss} = \lim_{s \to 0} \left[-s \frac{G_2(s)}{1 + G_2(s)[G_1(s) + 1]} \theta_d(s) \right] \qquad [22]$$

The total error when there is both a reference input and a disturbance is thus the sum of the errors given by equations [21] and [22].

Example 9

A liquid level control system, described in Example 9, Chapter 6, and by Fig. 6.23 and Fig. 6.24, has a system of the form shown in Fig. 7.12. What is the steady-state error when the system is subject to a step input disturbance of size A?

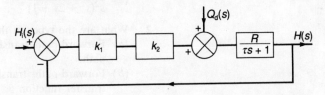

Fig. 7.12 Example 9

Answer

Figure 7.13(a) shows the system when there is no reference input and only a disturbance input. Figure 7.13(b) shows how this can be rearranged to give a unity feedback system. Thus for such a system the open-loop transfer function is

$$G_o(s) = \frac{R/(\tau s + 1)}{1 + [R/(\tau s + 1)](k_1 k_2 + 1)}$$

Hence the output $H(s)$ is related to the input $Q_d(s)$ by

$$H(s) = \frac{R/(\tau s + 1)}{1 + [R/(\tau s + 1)](k_1 k_2 + 1)} Q_d(s)$$

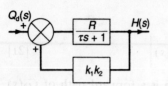

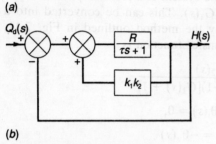

(a)

(b)

Fig. 7.13   Example 9

The error is thus, since $H_i(s) = 0$,

$$E(s) = H_i(s) - H(s) = -H(s)$$

$$E(s) = -\frac{R/(\tau s + 1)}{1 + [R/(\tau s + 1)](k_1 k_2 + 1)} Q_d(s)$$

Hence the steady-state error when there is a step input A/s is

$$e_{ss} = \lim_{s \to 0} \left\{ -s \frac{R/(\tau s + 1)}{1 + [R/(\tau s + 1)](k_1 k_2 + 1)} \frac{A}{s} \right\}$$

$$= \lim_{s \to 0} \left[\frac{RA}{\tau s + 1 + R k_1 k_2 + R} \right]$$

$$= \frac{RA}{1 + R k_1 k_2 + R}$$

Problems

1   State the type of the systems that gave the following open-loop transfer functions:

(a) $\dfrac{5}{s + 2}$

(b) $\dfrac{2(s + 1)}{s^2 + 2s + 1}$

(c) $\dfrac{6}{s(s + 3)}$

(d) $\dfrac{2}{s^2 + 4s}$

(e) $\dfrac{2(s + 3)}{s^2(s^2 + 2s + 1)}$

2   What are the types of the following closed-loop systems?
(a) Forward-path transfer function $1/(2s + 1)$, unity feedback path.
(b) Forward-path transfer function $1/(s + 1)$, feedback path transfer function 5.
(c) Forward-path transfer function $1/(s^2 + 2s + 1)$, feedback path transfer function $1/s$.
(d) Forward-path transfer function $2(s + 1)/(s^2 + 5s)$, feedback path transfer function $1/(s + 1)$.

3   Calculate the steady-state error for the system shown in Fig. 7.14 when subject to a unit step input if K has the value (a) 1, (b) 10 and comment on the significance of increasing the value of K.

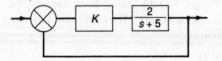

Fig. 7.14   Problem 3

4   The steering system of a car has an open-loop transfer function of

$$G_o(s) = \frac{K}{s(s + a)(s + b)}$$

What will be the steady-state errors when the steering is subject to (*a*) a step input of size A and (*b*) a ramp input changing at the rate A?

5   What are the steady-state errors for the systems giving the open-loop transfer functions listed in Problem 1 if they are subject to (i) a unit step input, (ii) a unit ramp input, (iii) a unit parabolic input?

6   Determine the steady-state error that will occur with a linear system having an open-loop transfer function of

$$\frac{10}{s(s+5)}$$

when subject to an input of

$$\frac{2}{s} - \frac{3}{s^2}$$

7   What are the type numbers for the system shown in Fig. 7.15 when subject to (*a*) the reference input and (*b*) the disturbance?

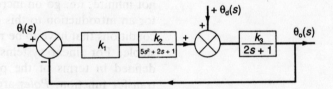

Fig. 7.15   Problem 7

8 Poles, zeros and stability

Introduction

An important requirement for a control system is that it should be stable. This means that if a finite sized input is applied to the system then the output should also be finite and not infinite, i.e. go on increasing without limit (see Chapter 1 for an introduction to this concept). This chapter is about the conditions that have to be realized for systems if they are to be stable. For linear systems the stability requirement can be defined in terms of the poles and zeros of the closed-loop transfer function. *Poles* are the roots of the transfer function numerator and *zeros* the roots of the denominator.

Defining stability

A system can be defined as *stable* if every bounded, i.e. finite, input produces a bounded output. Thus, for example, for every step input applied to a system the output must be finite. A system is not necessarily stable if a single step input results in a finite output: every step input must.

Alternatively, a system can be defined as stable if when it is subject to an impulse input the output dies away to zero as the time tends to infinity. If, following the impulse input, the system output tends to infinity as the time tends to infinity then the system is *unstable*. If, however, the output does not die away to zero or increase to infinity but tends to some finite but non-zero value then the system is said to be *critically* or *marginally stable*. Chapter 1 gives the example of a system involving a ball resting on a saucer-shaped surface. If the ball is inside the saucer then it is stable since after an impulse the ball would end up as time tended to infinity in the same position in the centre of the saucer from which it started. If the ball is on the convex upside-down surface of the saucer then the impulse would cause the ball to roll off and not return to its original position as time tended to infinity. Such a system is unstable. If the ball was resting on a flat surface then the impulse would result in the ball moving along the surface and, as time tended to infinity, coming to rest in a stable position some distance from its start point. Such a system is critically or marginally stable.

Example 1

Which of the following systems are stable?
(a) A step input to the system produces an output which can be described by the equation $\theta_o = 2t$.
(b) A step input to the system produces an output which can be described by the equation $\theta_o = 5$.
(c) An impulse applied to the system produces an output which can be described by the equation $\theta_o = e^{-t}$.
(d) An impulse applied to the system produces an output which can be described by the equation $\theta_o = e^t$.

Answer

(a) Figure 8.1(a) shows the form of the output. Since it goes on increasing and is not bounded the system is unstable.
(b) Figure 8.1(b) shows the form of the output. Since it is bounded, i.e. finite, the system is stable.
(c) Figure 8.1(c) shows the form of the output. Since it dies away with time, i.e. tends to zero as time tends to infinity, the system is stable.
(d) Figure 8.1(d) shows the form of the output. Since it continues increasing as time tends to infinity the system is unstable.

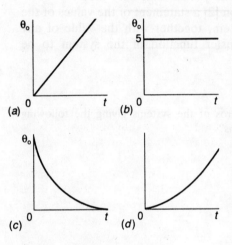

Fig. 8.1 Problem 1

Poles and zeros

The closed-loop transfer function $G(s)$ of a system can, in general, be represented by

$$G(s) = \frac{K(s^m + a_{m-1}s^{m-1} + a_{m-2}s^{m-2} + \ldots a_1 s + a_0)}{(s^n + b_{n-1}s^{n-1} + b_{n-2}s^{n-2} + \ldots b_1 s + b_0)} \quad [1]$$

and, if the roots of the denominator and numerator are established, as

$$G(s) = \frac{K(s + z_1)(s + z_2) \ldots (s + z_m)}{(s + p_1)(s + p_2) \ldots (s + p_n)} \quad [2]$$

where the roots of the numerator are $z_1, z_2, \ldots z_m$ and are called *zeros* and the roots of the denominator are $p_1, p_2, \ldots p_n$ and are called *poles*, K is a constant multiplier or gain of the system.

The zeros are the values of s for which the transfer function becomes zero. The poles are the values of s for which the transfer function is infinite, i.e. they make the value of the denominator become zero. Thus if the numerator is $(s - 2)$ then the transfer function is zero if $(s - 2) = 0$, i.e. $s = +2$. Hence the zero is $+2$. If the denominator is $(s + 5)$ then the transfer function is infinite if $(s + 5) = 0$, i.e. $s = -5$. Hence the pole is -5.

Zeros and poles can be real or complex quantities. Thus, for example, if the denominator was $(s^2 - 6s + 8)$ then since this is $(s - 4)(s - 2)$ the poles are $+4$ and $+2$. However, if the denominator was $(s^2 + 1)$ then since this can be written as $(s - 1j)(s + 1j)$ the poles are $+1j$ and $-1j$. In general, poles and zeros can be written as

$$s = \sigma + j\omega \qquad\qquad [3]$$

where σ is the real part and $j\omega$ the imaginary part.

As indicated by equation [2] a statement of the values of the zeros and poles of a system, together with the value of any gain K, enables the transfer function of the system to be specified completely.

Example 2

What are the poles and zeros of the systems giving the following closed-loop transfer functions?

(a) $\dfrac{s - 1}{s^2 - 4s + 4}$

(b) $\dfrac{2(s + 1)}{(s + 1)(s + 2)(s - 3)}$

(c) $\dfrac{(s + 3)(s - 1)}{s(s + 2)(s + 3)(s - 4)}$

(d) $\dfrac{s + 4}{s^2 + 1s + 3}$

(e) $\dfrac{1}{s^2 + s + 1}$

Answer

(a) The denominator can be written as $(s - 2)(s - 2)$ and so the poles are $+2$ and $+2$. The numerator is $(s - 1)$ and so the zero is $+1$.

(b) The denominator is $(s + 1)(s + 2)(s - 3)$ and so the poles are -1, -2, and $+3$. The numerator is $2(s + 1)$ and so the zero is -1.

(c) The denominator is $(s - 0)(s + 2)(s + 3)(s - 4)$ and so the poles are 0, -2, -3, $+4$. The numerator is $(s + 3)(s - 1)$ and so the zeros are -3, $+1$.

(d) The denominator is $(s^2 + 1s + 3)$. The roots can be obtained by using the equation for the roots m of a quadratic $(ax^2 + bx + c)$, i.e.

$$m = \frac{-b \pm \sqrt{(b^2 - 4ac)}}{2a}$$

Hence the poles of $(s^2 + 1s + 3)$ are

$$s = \frac{-1 \pm \sqrt{(1 - 12)}}{2}$$

$$= -0.5 \pm 0.5\sqrt{(-1)}\sqrt{(12-1)}$$

$$= -0.5 \pm j1.7$$

Hence the poles are $(-0.5 + j1.7)$ and $(-0.5 - j1.7)$. The numerator is $(s - 4)$ and so the zero is -4.

(e) The denominator is $(s^2 + s + 1)$. Using the expression given above for determining the roots

$$s = \frac{-1 \pm \sqrt{(1-4)}}{2}$$

$$= -0.5 \pm 0.5\sqrt{(-1)}\sqrt{(4-1)}$$

$$= -0.5 \pm j0.87$$

Hence the poles are $(-0.5 + j0.87)$ and $(-0.5 - j0.87)$. The numerator is 1, a constant, and hence there are no zeros.

Example 3

What are the transfer functions of the systems having the following poles and zeros?
(a) Poles -1, -2; no zero.
(b) Poles $+1$, -2; zero 0.
(c) Poles $(-2 \pm j1)$, zero $+1$.
(d) Poles $(1 \pm j2)$, zero -1.

Answer

(a) The denominator will be $(s + 1)(s + 2)$ and the numerator 1 (in the absence of any information about the gain K). Hence

$$G(s) = \frac{1}{(s + 1)(s + 2)}$$

(b) The denominator will be $(s - 1)(s + 2)$ and the numerator $(s - 0)$. Hence

$$G(s) = \frac{s}{(s - 1)(s + 2)}$$

(c) The denominator will be

$$[s - (-2 + j1)][s - (-2 - j1)]$$
$$= [s^2 - (-2 + j1)s - (-2 - j1)s + (-2 + j1)(-2 - j1)]$$
$$= [s^2 + 4s + 5]$$

The numerator is $(s - 1)$, and so

$$G(s) = \frac{s - 1}{s^2 + 4s + 5}$$

(d) The denominator will be

$$[s - (1 + j2)][s - (1 - j2)]$$
$$= [s^2 - (1 + j2)s - (1 - j2)s + (1 + j2)(1 - j2)]$$
$$= [s^2 - 2s + 5]$$

The numerator is $(s + 1)$ and so

$$G(s) = \frac{s + 1}{s^2 - 2s + 5}$$

Pole-zero plots

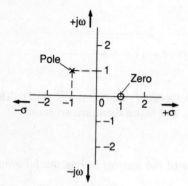

Fig. 8.2 Pole-zero plot

The poles and zeros of a transfer function can be represented on a diagram, called the *pole-zero plot*. Figure 8.2 shows the axes used for such a plot. The *x*-axis is the real part of the pole or zero, the *y*-axis the imaginary part. The position of a pole is marked with a cross '×' and the position of a zero with a small circle 'o'. Thus in the figure a pole is marked as having a real part of −1 and an imaginary part of +1, i.e. the pole is $(-1 + j1)$. The zero is marked as being +1, there being no imaginary part.

The two-dimensional plot is known as the *s-plane*. Poles or zeros in the left-hand side of the plot are all negative, poles or zeros in the right-hand side positive. Poles or zeros are either real or occur in pairs as $(\sigma \pm j\omega)$.

Example 4

Sketch the pole-zero plots for systems having the following poles and zeros.

(*a*) Poles −2, +3; zero +1.
(*b*) Poles 0, −1, −2; zero −3.
(*c*) Poles −1 ± j2; zero −1.
(*d*) Poles −2 ± j1, 0; zero −3 ± j2.

Answer

See Fig. 8.3.

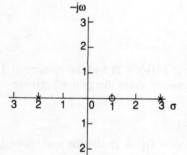

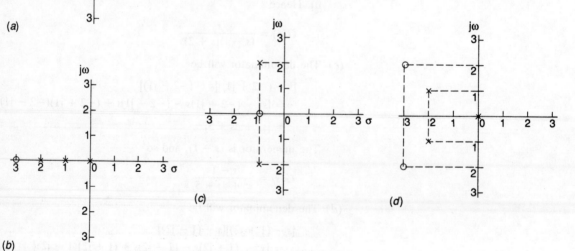

Fig. 8.3 Example 4

Stability and poles

The stability of a system can be determined by considering how the output changes with time after there has been an impulse input. With a stable system the output should die away to zero with time, with an unstable system the output will grow with time.

Consider a system with no zero and a pole of −2. The transfer function $G(s)$ will be

$$G(s) = \frac{1}{s+2}$$

Hence the output $\theta_o(s)$ is related to the input $\theta_i(s)$ by

$$\theta_o(s) = \frac{1}{s+2}\theta_i(s)$$

If the system is subject to a unit impulse then $\theta_i(s) = 1$ and so

$$\theta_o(s) = \frac{1}{s+2}$$

This is a Laplace transform of the form $1/(s+a)$ and so the inverse gives

$$\theta_o = e^{-2t}$$

The value of e^{-2t} decreases with time, becoming zero at infinite time. Thus the system is stable.

Now consider a system with no zero and a pole of +2. The transfer function $G(s)$ will be

$$G(s) = \frac{1}{s-2}$$

Hence

$$\theta_o(s) = \frac{1}{s-2}\theta_i(s)$$

For a unit impulse $\theta_i(s) = 1$ and so

$$\theta_o(s) = \frac{1}{s-2}$$

This is a Laplace transform of the form $1/(s+a)$ and so the inverse gives

$$\theta_o = e^{2t}$$

As t increases so the value of e^{2t} increases, hence the system is unstable.

Consider another example, this time with poles of $(-2 \pm j1)$ and no zero. The transfer function is

$$G(s) = \frac{1}{[s - (-2 + j1)][s - (-2 - j1)]}$$

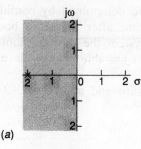

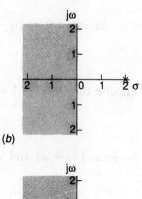

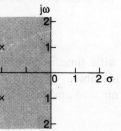

Fig. 8.4 Pole-zero plots and stability, stable region shaded. (a) Pole −2, no zero: stable. (b) Pole +2, no zero: unstable. (c) Poles (−2 ± j1), no zero: stable.

and so

$$\theta_o(s) = \frac{1}{[s - (-2 + j1)][s - (-2 - j1)]} \theta_i(s)$$

Thus when there is a unit impulse $\theta_i(s) = 1$, and so

$$\theta_o(s) = \frac{1}{[s - (-2 + j1)][s - (-2 - j1)]}$$

This is a Laplace transform of the form $1/[(s + a)(s + b)]$ and so the inverse is of the form

$$\frac{1}{b - a}(e^{-at} - e^{-bt})$$

with $a = -(-2 + j1)$ and $b = -(-2 - j1)$, and so

$$\theta_o = \frac{1}{j2}(e^{-2t}e^{jt} - e^{-2t}e^{-jt})$$

$$= e^{-2t}\left(\frac{e^{jt} - e^{-jt}}{j2}\right)$$

The term in the brackets is $\sin t$. Hence

$$\theta_o = e^{-2t}\sin t$$

This is a sine wave which has an amplitude which decays with time according to e^{-2t}. Thus the output decays with time and so the system is stable.

In general, when an impulse is applied to a system the output is in the form of the summation of a number of exponential terms. If just one of the exponential terms is of an exponential growth, i.e. the exponential of a positive function of t such as e^{2t}, then the output continually grows with time and the system is unstable. This situation will arise if any one of the poles has a real part which is positive and so the denominator of the transfer function includes a term $(s - a)$. When there are pole pairs involving $\pm j\omega$ then the output is always an oscillation. Such an oscillation is stable if the real part of the pole pair is negative and unstable if positive.

Thus if all the poles are in the left-hand side of the pole-zero plot then the system is stable. If just one pole is in the left-hand side it is unstable. A system is critically stable if one or more poles lie on the vertical axis of the pole-zero plot, i.e. has a zero real value, and no poles lie in the right-hand side. Figure 8.4 shows the pole positions for the above examples and Fig. 8.5 the general form that outputs take for different pole positions. It is only the poles of the system transfer function that are important as far as stability is concerned, the values of the system zeros are irrelevant.

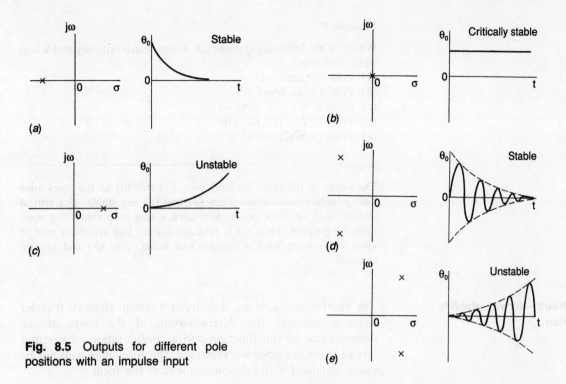

Fig. 8.5 Outputs for different pole positions with an impulse input

An alternative to the above form of analysis of stability is to consider the stability in terms of how the output changes with time after there has been a step input. This is a bounded input and for a stable system there should be a bounded output. The same conditions for stability in relation to pole positions results. Figure 8.6 shows the general form that outputs take for different pole positions.

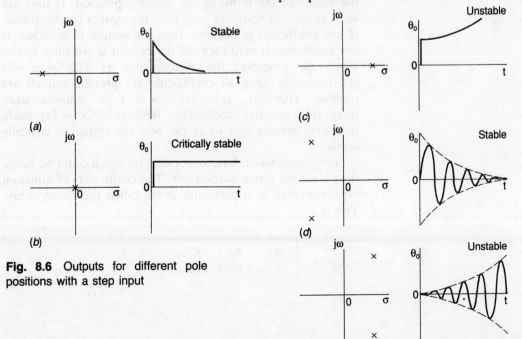

Fig. 8.6 Outputs for different pole positions with a step input

Example 5

Which of the following systems are stable, which critically stable and which unstable?

(a) Pole -4; zero $+1$.
(b) Pole $+1$; no zero.
(c) Poles 0, -1, -2; zero $+1$.
(d) Poles $(-2 \pm j3)$; no zero.
(e) Poles $(1 \pm j2)$; zero -2.

Answer

The values of the zeros are irrelevant. For stability all the poles must have negative real values. Thus (a) and (d) are stable. For critical stability one or more poles must have a real zero value and none must be positive. Thus (c) is critically stable. For instability one or more poles must have a positive real value, thus (b) and (e) are unstable.

The Routh–Hurwitz stability criterion

The determination of the stability of a system given its transfer function involves the determination of the roots of the denominator of the function and a consideration of whether any of them are positive. However, the roots may well not be easily obtained if the denominator is in the form

$$a_n s^n + a_{n-1} s^{n-1} + a_{n-2} s^{n-2} + \ldots a_1 s + a_0 \qquad [4]$$

and n is more than 3 or 4. The *Routh–Hurwitz criterion*, however, represents a method which can be used in such situations.

The first test that is applied is to inspect the coefficients, i.e. the values of the terms in the above expression. If they are all positive and none are zero then the system can be stable. If any coefficient is negative then the system is unstable. If any coefficient is zero then, at the best, it is critically stable. Thus, for example, the denominator $(s^3 + 2s^2 + 3s + 1)$ can be stable since all coefficients are present and all are positive. However, $(s^3 - 2s^2 + 3s + 1)$ is unstable since there is a negative coefficient. With $(s^3 + 2s^2 + 3s)$ there is a term missing and so at the best the system is critically stable.

For systems which have denominators which could be stable then a second test is carried out. The coefficients of equation [4] are written in a particular order called the *Routh array*. This is

s^n	a_n	a_{n-2}	a_{n-4}	$\ldots$
s^{n-1}	a_{n-1}	a_{n-3}	a_{n-5}	$\ldots$

Further rows in the array are determined by calculation from elements in the two rows immediately above. Successive rows are calculated until only zeros appear. The array should then contain $(n + 1)$ rows, a row corresponding to each of the terms s^n to s^0.

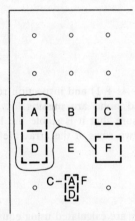

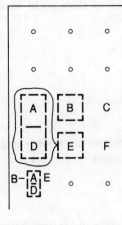

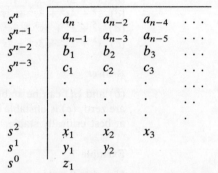

s^n	a_n	a_{n-2}	a_{n-4}	$\cdots$
s^{n-1}	a_{n-1}	a_{n-3}	a_{n-5}	$\cdots$
s^{n-2}	b_1	b_2	b_3	$\cdots$
s^{n-3}	c_1	c_2	c_3	$\cdots$
$\cdot$	$\cdot$	$\cdot$	$\cdot$	$\cdots$
$\cdot$	$\cdot$	$\cdot$	$\cdot$	$\cdots$
$\cdot$	$\cdot$	$\cdot$	$\cdot$	
s^2	x_1	x_2	x_3	
s^1	y_1	y_2		
s^0	z_1			

Elements in the third row are obtained from elements in the previous two rows by

$$b_1 = a_{n-2} - \left(\frac{a_n}{a_{n-1}}\right) a_{n-3} \qquad [5]$$

$$b_2 = a_{n-4} - \left(\frac{a_n}{a_{n-1}}\right) a_{n-5} \qquad [6]$$

Elements in the fourth row are obtained from elements in the previous two rows by

$$c_1 = a_{n-3} - \left(\frac{a_{n-1}}{b_1}\right) b_2 \qquad [7]$$

$$c_2 = a_{n-5} - \left(\frac{a_{n-1}}{b_1}\right) b_3 \qquad [8]$$

Fig. 8.7 Determining elements in the Routh array

One way of remembering these rules for determining the elements is illustrated in Fig. 8.7.

When the array has been completed it is inspected. If all the elements in the first column of the array are positive the roots all have negative real parts, and so lie in the left-hand side of the pole-zero plot. The system is thus stable if all the first-column elements are positive. If there are any negative elements in the first column the number of sign changes in the first column is equal to the number of roots with positive real parts.

Example 6

The following are the denominators of the transfer functions of a number of systems. By inspection, which of them could be stable, which unstable and which critically stable?

(a) $s^4 + 3s^3 + 2s + 3$.

(b) $s^3 + 2s^2 + 3s + 1$.

(c) $s^5 - 4s^4 + 3s^3 + 2s^2 + 5s + 2$.

(d) $s^5 + s^4 + 5s^3 + 2s^2 + 3s + 2$.

(e) $s^5 + 2s^3 + 3s^2 + 4s + 5$.

Answer

(b) and (d) can be stable since all coefficients are positive and none are zero. (c) is unstable since there is a negative term. (a) and (e) are at best critically stable.

Example 7

Use the Routh array to determine whether the system having the following transfer function is stable.

$$G(s) = \frac{2s + 1}{s^4 + 2s^3 + 3s^2 + 4s + 1}$$

Answer

The denominator is $(s^4 + 2s^3 + 3s^2 + 4s + 1)$ and inspection reveals that all coefficients are positive and none are missing. It could therefore be stable. In order to be sure that it is stable the Routh array has then to be used. The first two rows of the array are

$$
\begin{array}{c|ccc}
s^4 & 1 & 3 & 1 \\
s^3 & 2 & 4 &
\end{array}
$$

Elements in the third row of the array are calculated using equations [5] and [6].

$$b_1 = a_{n-2} - \left(\frac{a_n}{a_{n-1}}\right) a_{n-3}$$

$$b_1 = 3 - \left(\frac{1}{2}\right) 4 = 1$$

and

$$b_2 = a_{n-4} - \left(\frac{a_n}{a_{n-1}}\right) a_{n-5}$$

$$b_2 = 1 - \left(\frac{1}{2}\right) 0 = 1$$

Thus the array becomes

$$
\begin{array}{c|cc}
s^4 & 1 & 3 & 1 \\
s^3 & 2 & 4 \\
s^2 & 1 & 1
\end{array}
$$

Elements in the fourth row of the array are calculated using equations [7] and [8].

$$c_1 = a_{n-3} - \left(\frac{a_{n-1}}{b_1}\right) b_2$$

$$c_1 = 4 - \left(\frac{2}{1}\right) 1 = 2$$

and

$$c_2 = a_{n-5} - \left(\frac{a_{n-1}}{b_1}\right) b_3$$

$$c_2 = 0 - \left(\frac{2}{1}\right) 0 = 0$$

Thus the array becomes

s^4	1	3	1
s^3	2	4	
s^2	1	1	
s^1	2		

The fourth row element can be calculated using

$$d_1 = b_2 - \left(\frac{b_1}{c_1}\right) c_2$$

$$d_1 = 1 - \left(\frac{1}{2}\right) 0 = 1$$

Thus the array becomes

s^4	1	3	1
s^3	2	4	
s^2	1	1	
s^1	2		
s^0	1		

The first column has all positive elements and so the system is stable.

Example 8

Use the Routh array to determine whether the system having the following transfer function is stable:

$$G(s) = \frac{2s + 1}{s^4 + s^3 + s^2 + 4s + 1}$$

Answer

The denominator is $(s^4 + s^3 + s^2 + 4s + 1)$ and inspection reveals that all coefficients are positive and none are missing. It could therefore be stable. In order to be sure that it is stable the Routh array has then to be used. The first two rows of the array are

$$
\begin{array}{c|ccc}
s^4 & 1 & 1 & 1 \\
s^3 & 1 & 4 &
\end{array}
$$

Elements in the third row of the array are calculated using equations [5] and [6].

$$ b_1 = a_{n-2} - \left(\frac{a_n}{a_{n-1}}\right) a_{n-3} $$

$$ b_1 = 1 - \left(\frac{1}{1}\right) 4 = -3 $$

and

$$ b_2 = a_{n-4} - \left(\frac{a_n}{a_{n-1}}\right) a_{n-5} $$

$$ b_2 = 1 - \left(\frac{1}{1}\right) 0 = 1 $$

Thus the array becomes

$$
\begin{array}{c|ccc}
s^4 & 1 & 1 & 1 \\
s^3 & 1 & 4 & \\
s^2 & -3 & 1 &
\end{array}
$$

Elements in the fourth row of the array are calculated using equations [7] and [8].

$$ c_1 = a_{n-3} - \left(\frac{a_{n-1}}{b_1}\right) b_2 $$

$$ c_1 = 4 - \left(\frac{1}{-3}\right) 1 = \frac{13}{3} $$

and

$$ c_2 = a_{n-5} - \left(\frac{a_{n-1}}{b_1}\right) b_3 $$

$$ c_2 = 0 - \left(\frac{1}{-3}\right) 0 = 0 $$

Thus the array becomes

$$
\begin{array}{c|ccc}
s^4 & 1 & 1 & 1 \\
s^3 & 1 & 4 & \\
s^2 & -3 & 1 & \\
s^1 & 13/3 & &
\end{array}
$$

The fourth row element can be calculated using

$$ d_1 = b_2 - \left(\frac{b_1}{c_1}\right) c_2 $$

$$ d_1 = 1 - \left(\frac{-3}{13/3}\right) 0 = 1 $$

Thus the array becomes

$$
\begin{array}{c|ccc}
s^4 & 1 & 1 & 1 \\
s^3 & 1 & 4 & \\
s^2 & -3 & 1 & \\
s^1 & 13/3 & & \\
s^0 & 1 & &
\end{array}
$$

The first column has a negative element and so the system is unstable. There are two sign changes, plus to minus and then minus to plus, and so the system has two poles with positive real parts.

Example 9

The denominator of the transfer function of a system is

$$s^3 + 4s^2 + 8s + K$$

Within what range can K be if the system is to be stable?

Answer

The first two rows of the Routh array are

$$
\begin{array}{c|ccc}
s^3 & 1 & 8 & 1 \\
s^2 & 4 & K &
\end{array}
$$

Elements in the third row of the array are calculated using equations [5] and [6].

$$b_1 = a_{n-2} - \left(\frac{a_n}{a_{n-1}}\right)a_{n-3}$$

$$b_1 = 8 - \left(\frac{1}{4}\right)K$$

and

$$b_2 = 0 - \left(\frac{1}{4}\right)0 = 0$$

Hence the array becomes

$$
\begin{array}{c|cc}
s^3 & 1 & 8 \\
s^2 & 4 & K \\
s^1 & 8 - \tfrac{1}{4}K & 0
\end{array}
$$

The element in the fourth row of the array is calculated using equation [7].

$$c_1 = a_{n-3} - \left(\frac{a_{n-1}}{b_1}\right)b_2$$

$$c_1 = K - \left(\frac{4}{8 - \tfrac{1}{4}K}\right)0 = K$$

Hence the array becomes

$$
\begin{array}{c|cc}
s^3 & 1 & 8 \\
s^2 & 4 & K \\
s^1 & 8 - \frac{1}{4}K & 0 \\
s^0 & K &
\end{array}
$$

For the system to be stable all the elements in the first column must be positive. This means

$$(8 - \tfrac{1}{4}K) > 0$$

and

$$K > 0$$

This means $8 > \frac{1}{4}K$ and so $32 > K$. Hence K must be between 0 and 32.

Example 10

For the system shown in Fig. 8.8, what range of K will result in stability?

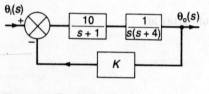

$\theta_i(s)$ $\dfrac{10}{s+1}$ $\dfrac{1}{s(s+4)}$ $\theta_o(s)$ K

Fig. 8.8 Example 10

Answer

The overall system transfer function is given by

$$\frac{G(s)}{1 + G(s)H(s)}$$

and so, since the forward path transfer function is $10/[s(s + 1)(s + 4)]$, is

$$\frac{10/[s(s + 1)(s + 4)]}{1 + 10K/[s(s + 1)(s + 4)]} = \frac{10}{s^3 + 5s + 4s + 10K}$$

Hence the Routh array for the denominator is

$$
\begin{array}{c|cc}
s^3 & 1 & 4 \\
s^2 & 5 & 10K \\
s^1 & 4 - 2K & \\
s^0 & 10K &
\end{array}
$$

For the first column to only have positive values we must have

$$4 - 2K > 0$$

and

$$10K > 0$$

This means that K must lie between 0 and 2.

Relative stability

Construction of the Routh array and application of the criterion that the first column should only contain positive

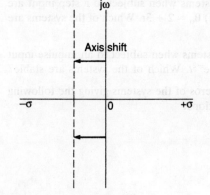

Fig. 8.9 Axis shifting

terms enables it to be decided whether a system has all its roots in the left-hand side of the *s* plane and so whether the system is stable. However, it is often useful to know for a stable system how close it is to being unstable, i.e. its relative stability. To do this we need to know how close the roots are to the zero axis. This can be done by shifting the axis to the left by some amount (Fig. 8.9) and finding out whether the shift results in an unstable system measured relative to the new axis.

The shift of the axis to $-\sigma$ means that in the denominator of the tranfer function all the values of s are replaced by $(r - \sigma)$, where the equation in r is now being tested for stability.

Example 11

Has the system having a transfer function with the following denominator any roots closer to the zero axis than -1?

$$s^3 + 4s^2 + 8s + 4$$

Answer

We can test whether the system is stable by constructing the Routh array. This is

$$
\begin{array}{c|cc}
s^3 & 1 & 8 \\
s^2 & 4 & 4 \\
s^1 & 7 & \\
s^0 & 4 & \\
\end{array}
$$

and so the system is stable. If we now shift the axis to -1 then we replace s by $(r - 1)$.

$$(r - 1)^3 + 4(r - 1)^2 + 8(r - 1) + 4$$

This is

$$(r^3 - 3r^2 + 3r - 1) + 4(r^2 - 2r + 1) + 8(r - 1) + 4$$

After simplification this becomes

$$r^3 + r^2 + 3r - 1$$

The Routh array for this equation is

$$
\begin{array}{c|cc}
r^3 & 1 & 3 \\
r^2 & 1 & -1 \\
r^1 & 4 & \\
r^0 & -1 & \\
\end{array}
$$

The system is unstable. Since there is just one change of sign there is just one root to the right of the -1 line.

Problems

1 The outputs of several systems when subject to a step input are (a) $\theta_o = 3$, (b) $\theta_o = 3t$, (c) $\theta_o = 2 + 3t$. Which of the systems are stable?

2 The outputs of several systems when subject to an impulse input are (a) $2t$, (b) e^{-2t}, (c) te^{-2t}. Which of the systems are stable?

3 What are the poles and zeros of the systems giving the following closed-loop transfer functions?

(a) $\dfrac{2s - 1}{s^2 - 5s + 6}$

(b) $\dfrac{4s - 1}{s^2 + 2s}$

(c) $\dfrac{(s + 2)(s - 3)}{(s + 2)(s + 3)(s - 4)}$

(d) $\dfrac{4}{s^2 + 2s + 3}$

(e) $\dfrac{s^2 + s + 1}{s(s - 3)(s + 5)}$

4 What are the transfer functions of the systems having the following poles and zeros?
(a) Poles 0, −1, −2, zeros none.
(b) Poles +1, −3, zero 0.
(c) Poles $(-2 \pm j1)$, zero −1.
(d) Poles 0, $(3 \pm j2)$, zeros −1, +2.

5 Which of the following systems are stable, which critically stable and which unstable?
(a) Poles 0, −1, zeros none.
(b) Poles 0, +1, zero −1.
(c) Poles +1, −3, −4, zero −1.
(d) Poles −2, −3, zero 0.
(e) Poles $(-3 \pm j1)$, zero 0.
(f) Poles $(2 \pm j3)$, zero −1.
(g) Poles 0, $(-1 \pm j2)$, zeros +1, −2.

6 The following are the denominators of the transfer functions of a number of systems. By inspection, which of them could be stable, which unstable and which critically stable?
(a) $s^2 + 2s$.
(b) $s^3 - 2s + s + 3$.
(c) $s^4 + 3s^3 + 2s^2 + 4s - 1$.
(d) $s^5 + 2s^4 + 3s^3 + s^2 + 2s + 1$.

7 Use the Routh array to determine whether the systems having transfer function with the following denominators are stable.
(a) $s^3 + 4s + 8s + 12$.
(b) $s^4 + s^3 + s^2 + 2s + 3$.
(c) $s^3 + 2s^2 + 2s + 6$.
(d) $s^4 + 4s^3 + 24s^2 + 32s + 16$.

8 The denominator of the transfer function of a system is

$$s^3 + 2s^2 + 4s + K.$$

Within what range can K be if the system is to be stable?

9 Has the system having a transfer function with the following denominator any roots closer to the zero axis than -1?

$$s^2 + 4s + 2$$

10 Determine the range of values of K for the system designed by Fig. 8.10 which will result in stability if disturbances occur.

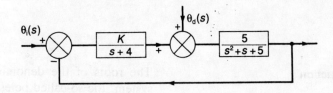

Fig. 8.10 Problem 10

9 Root locus analysis

Introduction

The roots of the denominator of the transfer function of a system, the so-called poles, determine the general form of the transient response of that system. Figure 9.1 shows, for a system having a transfer function

$$G(s) = \frac{K}{(s + p_1)(s + p_2)}$$

how changing the position of the poles p_1 and p_2 in the s-plane changes the transient response when the system is subject to an impulse. This chapter is concerned with a study of this relationship between behaviour of systems and the positions of their roots. The technique used for this study is called the *root locus method*.

First-order system root loci

Consider the first-order system shown in Fig. 9.2. The open-loop transfer function of the system $G_o(s)$ is $K/(s + 1)$ and, since the feedback is unity, the system has a transfer function $G(s)$ of

$$G(s) = \frac{K/(s + 1)}{1 + (K/(s + 1))}$$

which can be written as

$$G(s) = \frac{K}{s + (1 + K)}$$

The system has a single pole, at $-(1 + K)$. When $K = 0$ then the pole is -1 and as the value of K increases so the value of the pole becomes more negative, as shown in Fig. 9.3. The line showing how the position of the pole changes as K changes as it moves away from the $K = 0$ pole is called the *root locus*.

When $K = 0$ the transfer function of the system becomes the open-loop transfer function and so the value of the root for the

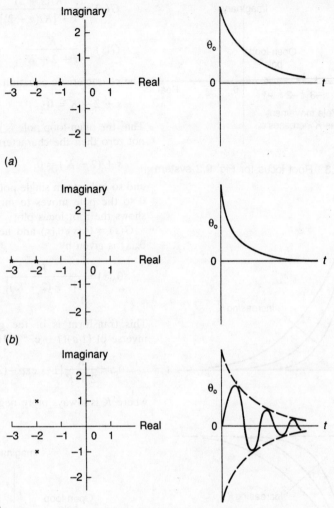

Fig. 9.1 Pole positions and responses to an impulse

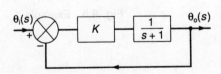

Fig. 9.2 A first-order system

system when $K = 0$ is called the *open-loop pole*.

Since the value of the root depends on the value of K so the response of the system depends on the value of K. Figure 9.4 shows how the response of the system to (*a*) an impulse and (*b*) a step depends on the value of K.

Example 1

For a first-order system having an open-loop transfer function of $K/(s + 2)$ and unity feedback, (*a*) sketch the root locus plot and (*b*) give the general response of the system to a unit step input.

Answer

The system has a transfer function of

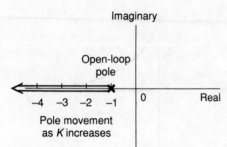

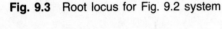

Fig. 9.3 Root locus for Fig. 9.2 system

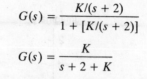

$$G(s) = \frac{K/(s+2)}{1 + [K/(s+2)]}$$

$$G(s) = \frac{K}{s+2+K}$$

The characteristic equation is

$$s + 2 + K = 0$$

Thus the open-loop pole is when $K = 0$ and so at $s = -2$. When K is not zero then the characteristic equation can be written as

$$s + (2 - K) = 0$$

and so indicates a single pole at $-(2 + K)$. Thus as K increases from 0 so the pole moves to more negative values from -2. Figure 9.5 shows the root locus plot.

$G(s) = \theta_o(s)/\theta_i(s)$ and hence for an input $\theta_i(s)$ of $1/s$ the output $\theta_o(s)$ is given by

$$\theta_o(s) = \frac{K}{s[s + (2 + K)]}$$

This transform is of the general form $1/[s(s + a)]$. This has the inverse of $(1/a)(1 - e^{-at})$. Hence

$$\theta_o = \frac{K}{2 + K}[1 - \exp-(2 + K)t]$$

where K is always more negative than -2.

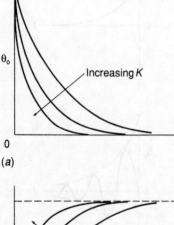

(a)

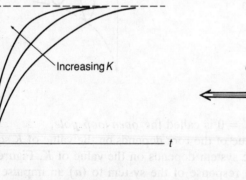

(b)

Fig. 9.4 Response of Fig. 9.2 system to (a) an impulse, (b) a step input

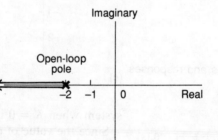

Fig. 9.5 Example 1

Second-order system root loci

Figure 9.6 shows a second-order system. The open-loop transfer function $G_o(s)$ of the system is $K/[s(s + 1)]$ and, since the feedback is unity, the system has a transfer function of

$$G(s) = \frac{K/[s(s + 1)]}{1 + K/[s(s + 1)]}$$

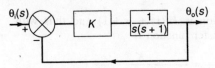

Fig. 9.6 A second-order system

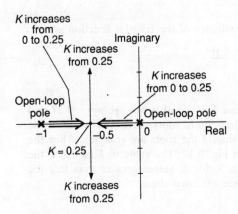

Fig. 9.7 Root loci for Fig. 9.6 system

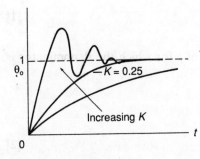

Fig. 9.8 Response of Fig. 9.6 system to a unit step input

which can be written as

$$G(s) = \frac{K}{s^2 + s + K}$$

The roots of an equation of the form $ax^2 + bx + c$ are given by

$$\text{Roots} = \frac{-b \pm \sqrt{(b^2 - 4ac)}}{2a}$$

and so the roots of the denominator of the transfer function are

$$p = \frac{-1 \pm \sqrt{(1 - 4K)}}{2}$$

$$p = -\tfrac{1}{2} \pm \tfrac{1}{2}\sqrt{(1 - 4K)}$$

When $K = 0$ then $p = -\tfrac{1}{2} \pm \tfrac{1}{2}$, i.e. the open-loop roots are at 0 and -1. When $K = \tfrac{1}{4}$ then $p = -\tfrac{1}{2}$, i.e. the roots are both at $-\tfrac{1}{2}$. For values of K between 0 and $\tfrac{1}{4}$ the root at 0 is becoming more negative and moving towards $-\tfrac{1}{2}$ while the root at -1 is becoming less negative and moving towards $-\tfrac{1}{2}$, as shown in Fig. 9.7. For $K = 1$ the roots are given by $p = -\tfrac{1}{2} \pm \sqrt{(-3)}$ and are thus $-\tfrac{1}{2} + j3$ and $-\tfrac{1}{2} - j3$. For all values of K greater than 0.25 a complex pair of roots occurs, with the real component being constant as $-\tfrac{1}{2}$ and the imaginary part having a value that increases as K increases.

Since the values of the roots depend on the value of K so the response of the system to inputs depends on the value of K. Figure 9.8 shows the response of the system with different K values to a unit step input. For values of K between 0 and 0.25 the system is the overdamped response of a second order system. For $K = 0.25$ the system is critically damped. For K greater than 0.25 the system is underdamped and oscillations occur.

The denominator of the transfer function, i.e. $(s^2 + s + K)$, can be written in the form $(s^2 + 2\omega_n\zeta s + \omega_n^2)$, see Chapters 2 and 3, where ω_n is the natural angular frequency of oscillation and ζ the damping ratio. Thus for this system $\omega_n^2 = K$ and $2\omega_n\zeta = 1$ and so $\omega_n = \sqrt{K}$ and $\omega_n\zeta = \tfrac{1}{2}$, i.e. the values of the poles at critical damping, and $\zeta = 1/(2\sqrt{K})$. Thus as K increases from 0.25 the natural frequency increases and the damping ratio decreases.

Example 2

For the system shown in Fig. 9.9, sketch the root loci diagram and specify the value of gain K at which the system starts to oscillate.

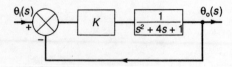

Fig. 9.9 Example 2

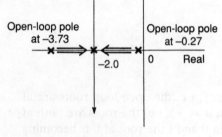

Open-loop pole at −3.73

Open-loop pole at −0.27

−2.0

Real

Fig. 9.10 Example 2

Root loci of closed-loop systems

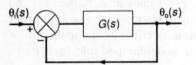

Fig. 9.11 Closed-loop system

Answer

The open-loop transfer function is $K/(s^2 + 4s + 1)$ and thus with unity feedback the system transfer function is

$$G(s) = \frac{K/(s^2 + 4s + 1)}{1 + K/(s^2 + 4s + 1)} = \frac{K}{s^2 + 4s + (1 + K)}$$

The roots of an equation of the form $ax^2 + bx + c$ are given by

$$\text{Roots} = \frac{-b \pm \sqrt{(b^2 - 4ac)}}{2a}$$

and so the roots of the denominator of the transfer function are

$$p = \frac{-4 \pm \sqrt{[16 - 4(1 + K)]}}{2}$$

$$p = -2 \pm \sqrt{(3 - K)}$$

The open-loop poles are when $K = 0$ and hence are $p = -0.27$ and $p = -3.73$. When $K = 3$ then both roots have the same value $p = -2$. When K is greater than 3 the roots are complex. Thus the root locus plot is as shown in Fig. 9.10. The value of K at which the system is critically damped is 3, for K values greater than this the roots are complex and so there are oscillations.

Consider the general closed-loop system shown in Fig. 9.11. The open-loop transfer function is $G_o(s)$ and thus, with unity feedback, the transfer function $G(s)$ for the system is

$$G(s) = \frac{G_o(s)}{1 + G(s)} \qquad [1]$$

The poles will be the values of s for which the denominator is zero, i.e.

$$1 + G_o(s) = 0$$

and so

$$G_o(s) = -1 \qquad [2]$$

$G_o(s)$ may derive from a grouping of a number of elements, each with its own transfer function. In general it is possible to write

$$G_o(s) = \frac{K(s - z_1)(s - z_2) \ldots (s - z_m)}{(s - p_1)(s - p_2) \ldots (s - p_n)} \qquad [3]$$

where K is a constant, $z_1, z_2, \ldots, z_m$ are the zeros and the $p_1, p_2, \ldots, p_n$ are the poles. If the values of s in the above equation are to be pole values and so lie on the root locus then equation [2] must also hold. Thus for points on the root locus

$$\frac{K(s - z_1)(s - z_2) \ldots (s - z_m)}{(s - p_1)(s - p_2) \ldots (s - p_n)} = -1 \qquad [4]$$

Because s is a complex variable the above equation may be written in polar form (see the note that follows for a brief discussion of such forms). Since the magnitude of the product of two complex numbers is the product of their magnitudes and the quotient is the quotient of their magnitudes then equation [4] in polar form gives for magnitudes

$$\frac{K|s - z_1||s - z_2| \ldots |s - z_\mathrm{m}|}{|s - p_1||s - p_2| \ldots |s - p_\mathrm{n}|} = 1 \qquad [5]$$

Since the argument of the product of two complex numbers is the sum of their arguments and the quotient their difference, then equation [4] in polar form gives for arguments

$$[\angle (s - z_1) + \angle (s - z_2) + \ldots + \angle (s - z_\mathrm{m})]$$
$$-[\angle (s - p_1) + \angle (s - p_2) + \ldots + \angle (s - p_3)]$$
$$= \pm \text{ odd multiple of } \pi \qquad [6]$$

Equation [6] can be used to determine whether a point in the s-plane is on a root locus. If it is on a root locus equation [6] will hold, if not on the root locus then it will not. By trial and error the root locus can then be established. Equation [5] will give the value of K at points along a root locus.

To illustrate this consider a system with an open-loop transfer function of

$$G_\mathrm{o}(s) = \frac{K(s + 1)}{s(s + 2)}$$

and unity feedback. The system transfer function will be

$$G(s) = \frac{K(s + 1)/[s(s + 2)]}{1 + K(s + 1)/[s(s + 2)]}$$

$$G(s) = \frac{K(s + 1)}{s(s + 2) + K(s + 1)}$$

The systems has open-loop poles, i.e. where $K = 0$, at 0 and -2 and a zero at -1. Consider some point on the s-plane of s_1 (Fig. 9.12). Lines are drawn connecting the point to the poles and zero. For s_1 to be on a root locus we must have, applying equation [6],

$$\beta - (\alpha_1 + \alpha_2) = \pm \text{ odd multiple of } \pi$$

Applying equation [5]

$$\frac{Kb}{ac} = 1$$

and so $K = ac/b$.

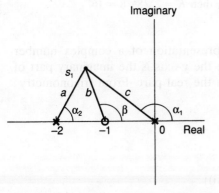

Fig. 9.12 $G_\mathrm{o}(s) = K(s + 1)/s(s + 2)$

As a further illustration, consider a system with

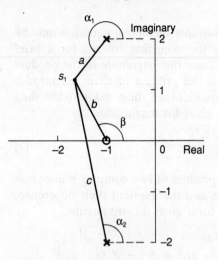

Fig. 9.13 $G_o(s) = K(s + 1)/(s^2 + 2s + 5)$

$$G_o(s) = \frac{K(s + 1)}{s^2 + 2s + 5}$$

and unity feedback. The transfer function of the system will be

$$G(s) = \frac{K(s + 1)}{s^2 + 2s + 5 + K(s + 1)}$$

This has a zero of -1 and open-loop roots, i.e. when $K = 0$, of $-1 \pm j2$ and hence the s-plane is as shown in Fig. 9.13. For point s_1 to be on a root locus we must have, applying equation [6]

$$\beta - (\alpha_1 + \alpha_2) = \pm \text{ odd multiple of } \pi$$

Applying equation [5]

$$\frac{Kb}{ac} = 1$$

and so $K = ac/b$.

Example 3

Show that the point s_1 in Fig. 9.14 lies on a root locus and determine the value of K for the point.

Answer

The system has three roots and no zero. Thus, applying equation [6]

$$0 - (\alpha_1 + \alpha_2 + \alpha_3) = \pm \text{ odd multiple of } \pi$$

But $\alpha_3 = 180°$ or π and $(\alpha_1 + \alpha_2) = 360°$ or 2π. Hence we have

$$0 - (\pi + 2\pi)$$

for the sum and so an odd multiple of π.

The value of K for the point is given by equation [5] as

$$\frac{K}{abc} = 1$$

Since $a = b = \sqrt{8}$ and $c = 2$, then $K = 2\sqrt{8}\sqrt{8} = 16$.

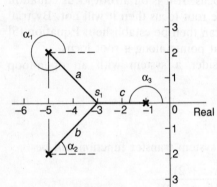

Fig. 9.14 Example 3

Polar representation of complex numbers

Figure 9.15 shows the representation of a complex number $(x + jy)$ on a graph where the y-axis is the imaginary part of the number and the x-axis the real part. From trigonometry

$$x = r\cos\theta$$
$$y = r\sin\theta$$

Thus

$$x + jy = r(\cos\theta + j\sin\theta)$$

This is usually abbreviated to

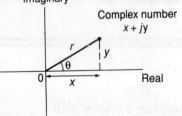

Fig. 9.15 Polar form of a complex number $x + jy = r\angle\theta$

r is called the *magnitude* or *modulus* of the complex number and is written as $|x + jy|$. θ is called the *argument*.

The product of two complex numbers can thus be written as

$$r_1(\cos\theta_1 + j\sin\theta_1)r_2(\cos\theta_2 + j\sin\theta_2)$$
$$= r_1 r_2[(\cos\theta_1\cos\theta_2 - \sin\theta_1\sin\theta_2)$$
$$+ j(\sin\theta_1\cos\theta_2 + \cos\theta_1\sin\theta_2)]$$
$$= r_1 r_2[\cos(\theta_1 + \theta_2) + j\sin(\theta_1 + \theta_2)]$$

and so the product has a magnitude which is the product of the magnitudes and an argument which is the sum of the arguments.

Dividing one complex number by another gives

$$\frac{r_1(\cos\theta_1 + j\sin\theta_1)}{r_2(\cos\theta_2 + j\sin\theta_2)}$$

$$= \frac{r_1(\cos\theta_1 + j\sin\theta_1)(\cos\theta_2 - j\sin\theta_2)}{r_2(\cos\theta_2 + j\sin\theta_2)(\cos\theta_2 - j\sin\theta_2)}$$

$$= \frac{r_1}{r_2}\left[\frac{(\cos\theta_1\cos\theta_2 + \sin\theta_1\sin\theta_2)}{\cos^2\theta_2 + \sin^2\theta_2}\right.$$

$$\left. + \frac{j(\sin\theta_1\cos\theta_2 - \cos\theta_1\sin\theta_2)}{\cos^2\theta_2 + \sin^2\theta_2}\right]$$

$$= \frac{r_1}{r_2}[\cos(\theta_1 - \theta_2) + j\sin(\theta_1 - \theta_2)]$$

and so the quotient has a magnitude which is the quotient of the magnitudes and an argument which is the difference of the arguments.

Construction of root loci

The technique outlined earlier in this chapter for the plotting of a root locus can be summarized as:

1 Find by trial and error those points in the s-plane which have angles between the real axis and the lines joining them to the zeros and poles which satisfy equation [6], the argument equation.

2 Determine the value of K at points on the root locus using equation [7], the magnitude equation.

This would appear to be rather a hit or miss procedure, however there are a number of rules which aid significantly the chance of picking on root locus points.

1 The number of root loci is equal to the order n of the open-loop transfer-function characteristic equation, i.e. the denominator. Each root traces out a locus as K varies from 0 at an open-loop pole to infinity at a zero. The term

root locus branch is often used for each root locus, the branches being continuous curves that start at each of the *n* open-loop poles, where $K = 0$, and approaches infinity at the *m* open-loop zeros. Locus branches for excess poles, i.e. where $n > m$, extend to infinity. For excess zeros, i.e. $m > n$, the branches extend from infinity to the open-loop poles. Thus for a system having an open-loop characteristic equation of $s^3 + 2s^2 + 3s + K = 0$ there will be three root locus branches.

2 The root loci of a system with a real characteristic equation are symmetrical with respect to the real axis. This is because complex roots occur in pairs of the form $\sigma \pm j\omega$.

3 The root loci start at the *n* poles of the system where $K = 0$.

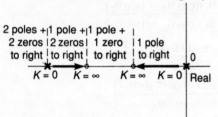

4 The root loci end at the *m* zeros of the system, where $K = \infty$. If there are more poles than zeros, the usual case, then *m* loci end at the *m* zeros and the remaining $(n - m)$ loci end at infinity.

5 Portions of the real axis are sections of root loci if the number of poles and zeros lying on the axis to the right of the portion is odd. Figure 9.16 illustrates this.

Fig. 9.16 Root loci on the real axis

6 Those loci terminating at infinity tend towards asymptotes at angles to the positive real axis of

$$\frac{\pi}{n - m}, \quad \frac{3\pi}{n - m}, \quad \frac{5\pi}{n - m}, \quad \ldots, \quad \frac{[2(n - m) - 1]\pi}{n - m}$$

Figure 9.17 shows examples of such loci for a system where $n = 3$ and $m = 0$. The angles of the asymptotes are $\pi/3$ or $60°$, π or $180°$ and $5\pi/3$ or $300°$.

7 The asymptotes intersect on the real axis at a point, sometimes called the *centre of gravity* or *centroid* of the asymptotes, given by

$$\frac{(p_1 + p_2 + \ldots p_n) - (z_1 + z_2 + \ldots + z_m)}{n - m}$$

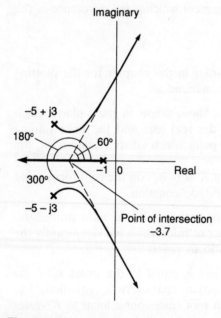

Thus for the example given in Fig. 9.17 where the poles are -1 and $-5 \pm j3$ and there are no zeros then the point of intersection is

$$\frac{-1 - 5 + j3 - 5 - j3}{3} = -3.7$$

Fig. 9.17 Asymptotes with $n = 3$ and $m = 0$

8 The intersection of root loci with the imaginary axis can be found by calculating those values of K which result in the existence of imaginary characteristic roots, i.e. $s = \sigma + j\omega$

with $\sigma = 0$. For example, with a characteristic equation $s^3 + 2s^2 + 3s + K$, then putting $s = j\omega$ gives $-j\omega^3 - 2\omega^2 + 3j\omega + K = 0$ and so equating imaginary parts gives $-\omega^3 + 3\omega = 0$ and $\omega = \sqrt{3}$ and equating real parts gives $-2\omega^2 + K = 0$ and so $K = 6$.

An alternative way of determining this intersection is to use the Routh array and find the limiting value of K that gives stability, this being the value of K where the root loci cross the imaginary axis.

9 The term *breakaway point* is used for where two or more loci meet at a point and subsequently 'break away' from that point along separate paths. Figure 9.18 shows such a breakaway point. Breakaway points occur at those points for which, for the characteristic equation, $dK/ds = 0$. However, it should be noted that not all roots of the equation $dK/ds = 0$ correspond to breakaway points, only those for which equation [5], i.e. the argument equation, holds. For the system giving Figure 9.18

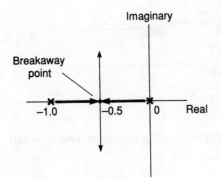

Fig. 9.18 Breakaway point
$G_o(s) = K/s(\sigma + 1)$

$$G(s) = \frac{K}{s^2 + s + K}$$

The characteristic equation is

$$s^2 + s + K = 0$$

Hence

$$K = -s^2 - s$$

$$\frac{dK}{ds} = -2s - 1$$

When $dK/ds = 0$ then

$$-2s - 1 = 0$$

and so the breakaway point is at $s = -\frac{1}{2}$.

10 The angle of departure of the locus at $K = 0$ from a complex pole and the angle of arrival of the locus at $K = \infty$ at a complex zero can be determined using the argument equation [6] and letting s be a point on the root locus very close to the pole or zero under consideration.

Figure 9.19 illustrates this being applied to determine the angle of departure from a complex pole $(2 + j2)$. The system has poles at 0 and $2 \pm j2$ and no zeros and so we must have

$$-(\alpha_1 + \alpha_2 + \alpha_3) = \pm \text{odd multiple of } \pi$$

Since the point is very close to the pole then $\alpha_1 = 180° - \tan^{-1}(2/2)$ and so is 135°. Angle α_3 is 90°. Thus we have

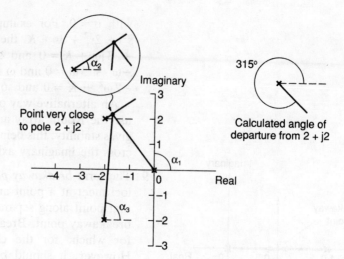

Fig. 9.19 Angle of departure

$$-(135° + \alpha_2 + 90°) = -\text{ odd multiple of } 180° = -540°$$

Hence $\alpha_2 = 315°$.

A useful sequence of steps that can be used, with the aid of the above rules, to construct root loci can be summarized as:

1 Obtain the characteristic equation of the open-loop transfer function $G_o(s)$ for the system.
2 Determine the positions of the poles and zeros.
3 Determine, using rule 1, the number of loci.
4 Plot the loci on the real axis using rule 5.
5 Determine the angles of the asymptotes using rule 6.
6 Obtain the intersection of the asymptotes with the real axis using rule 7.
7 Determine the intersection of the asymptotes with the imaginary axis using rule 8.
8 Determine the breakaway points using rule 9.
9 Obtain the angles of departures from complex poles and the angles of arrival at complex zeros using rule 10.
10 Sketch the root loci bearing in mind rules 2, 3 and 4.

Figure 9.20 shows some examples of root locus plots for systems having different forms of open-loop transfer function.

Example 4

Sketch the root loci for a system with an open-loop transfer function of

$$\frac{K}{(s + 1)(s + 2)(s + 3)}$$

Answer

Following the steps outlined above:

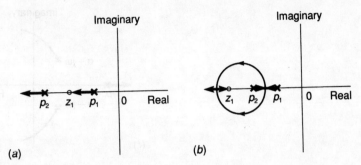

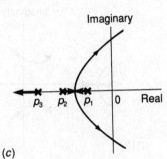

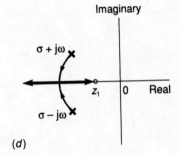

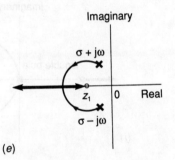

Fig. 9.20   Root locus plots:
(a) $G_o(s) = K(s + z_1)/(s + p_1)(s + p_2)$, when $p_2 > z_1 > p_1$;
(b) $G_o(s) = K(s + z_1)/(s + p_1)(s + p_2)$, when $z_1 > p_2 > p_1$;
(c) $G_o(s) = K/(s + p_1)(s + p_2)(s + p_3)$; (d) $G_o(s) = K(s + z_1)/$
$(s + \sigma + j\omega)(s + \sigma - j\omega)$, when $\sigma > z_1$; (e) $G_o(s) = K(s + z_1)/$
$(s + \sigma + j\omega)(s + \sigma - j\omega)$, when $z_1 > \sigma_1$; (f) $G_o(s) = K/(s + p_1)$
$(s + \sigma + j\omega)(s + \sigma - j\omega)$, when $p_1 > \sigma$; (g) $G_o(s) = K/(s + p_1)$
$(s + \sigma + j\omega)(s + \sigma - j\omega)$, when $\sigma > p_1$; (h) $G_o(s) = K(s + z_1)/$
$(s + p_1)(s + p_2)(s + \sigma + j\omega)(s + \sigma - j\omega)$, when $p_2 > p_1 > \sigma$;
(i) $G_o(s) = K(s + z_1)/s(s + p_1)(s + p_2)$, when $p_2 > z_1 > p_1$;
(j) $G_o(s) = K/s^2(s + p_1)$; (k) $G_o(s) = K(s + z_1)/s^2(s + p_1)$, when
$p_1 > z_1$; (l) $G_o(s) = K(s + z_1)/s^2(s + p_1)(s + p_2)$, when $p_2 > p_1 > z_1$.

(Figure continued on page 212)

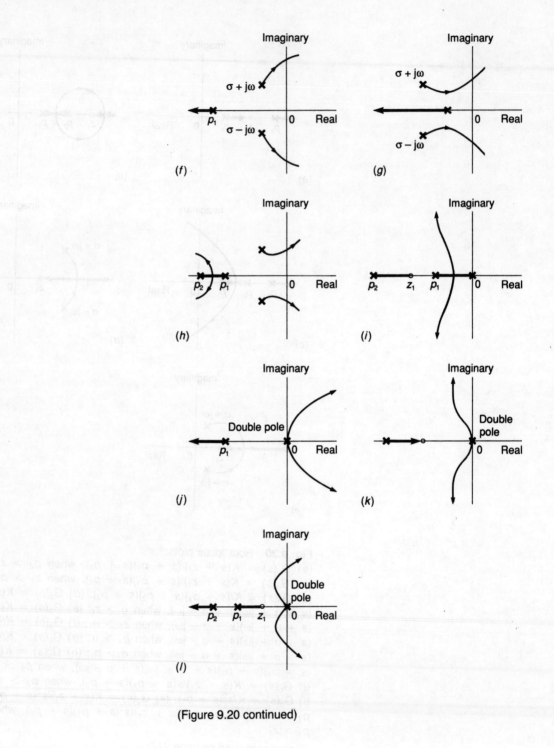

(Figure 9.20 continued)

1 The system has unity feedback and thus the system transfer function will be given by equation [1] as

$$G(s) = \frac{K/(s + 1)(s + 2)(s + 3)}{1 + K/(s + 1)(s + 2)(s + 3)}$$

$$G(s) = \frac{K}{(s + 1)(s + 2)(s + 3) + K}$$

The characteristic equation is thus

$$(s + 1)(s + 2)(s + 3) + K = 0$$

2 When $K = 0$ the characteristic equation becomes

$$(s + 1)(s + 2)(s + 3) = 0$$

and thus the open-loop poles are at -1, -2, and -3. There are no zeros.

3 The equation is third order and so there will be three root loci.

4 The portions of the root loci on the real axis will be between -1 and -2 and from -3 to infinity.

5 The angles of the asymptotes will be $\pi/3$ or $60°$, $3\pi/3$ or $180°$, and $5\pi/3$ or $300°$.

6 The point of intersection of the asymptotes with the real axis is

$$\frac{-1 - 2 - 3}{3} = -2$$

7 The intercepts with the imaginary axis can be determined by putting $s = j\omega$ in the characteristic equation which gives

$$(j\omega + 1)(j\omega + 2)(j\omega + 3) + K = 0$$

$$j^3\omega^3 + j^2\omega^2 6 + j\omega 11 + 6 + K = 0$$

$$-j\omega^3 - \omega^2 6 + j\omega 11 + 6 + K = 0$$

Equating the imaginary parts gives $-\omega^3 + 11\omega = 0$ and so $\omega = \pm \sqrt{11}$. Equating the real parts gives $-6\omega^2 + 6 + K = 0$ and so $K = 60$.

8 To determine the breakaway point: the characteristic equation is

$$(s + 1)(s + 2)(s + 3) + K = 0$$

$$K = -(s^3 + 6s^2 + 11s + 6)$$

$$\frac{dK}{ds} = -3s^2 - 12s - 11$$

Equating this to zero gives

$$3s^2 + 12s + 11 = 0$$

and so the roots are

$$\frac{-12 \pm \sqrt{(144 - 132)}}{6} = -2 \pm 0.58$$

Only the breakaway point $-2 + 0.58 = -1.42$ is feasible.

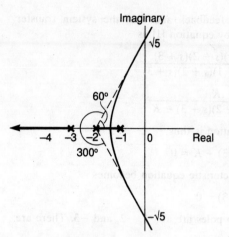

Fig. 9.21 Example 4

9 There are no complex poles or zeros.

10 Figure 9.21 shows the completed root loci plot.

Example 5

Sketch the root loci for a system with an open-loop transfer function of

$$\frac{K(s + 1)}{(s + 2 + j3)(s + 2 - j3)}$$

Answer

Following the steps outlined above:

1 The system has unity feedback and thus the system transfer function is

$$G(s) = \frac{K(s + 1)/(s + 2 + j3)(s + 2 - j3)}{1 + K(s + 1)/(s + 2 + j3)(s + 2 - j3)}$$

$$G(s) = \frac{K(s + 1)}{(s + 2 + j3)(s + 2 + j3) + K(s + 1)}$$

The characteristic equation is thus

$$(s + 2 + j3)(s + 2 - j3) + K(s - 1) = 0$$

2 When $K = 0$ the characteristic equation becomes

$$(s + 2 + j3)(s + 2 - j3) = 0$$

The open-loop poles are thus $2 \pm j3$. As the open-loop transfer function includes $(s + 1)$ in the numerator there is a zero of -1.

3 The equation is of second order and so there are two loci.

4 The portions of the root loci on the real axis are from -1 to infinity.

5 The angles of the asymptotes will be, since $n - m = 1$, $\pi/1$ or 180° and they are thus the real axis.

6 Since the asymptotes are the real axis the point of intersection has no significance.

7 Any points of intersection of loci with the imaginary axes can be determined by putting $s = j\omega$ in the characteristic equation. This gives

$$(j\omega + 2 + j3)(j\omega + 2 - j3) + K(j\omega - 1) = 0$$

$$-\omega^2 + j(4 + K)\omega + 13 - K = 0$$

Hence, equating imaginary parts gives $K = -4$ and equating real parts $\omega^2 = 13 - K$ and so $\omega = \pm 4.1$. When K has negative values the resulting loci are termed complementary loci. In this book the discussion is restricted to the loci produced when K is positive and thus the negative value of K indicates that the loci do not cross the imaginary axis.

8 The breakaway point is determined from the characteristic equation, which is

$$(s + 2 + j3)(s + 2 - j3) + K(s - 1) = 0$$

This can be simplified to

$$s^2 + 4s + 9 + K(s - 1) = 0$$

$$K = \frac{s^2 + 4s + 9}{1 - s}$$

Since for differentiation of a quotient we have

$$\frac{d}{dx}\left(\frac{u}{v}\right) = \frac{v(du/dx) - u(dv/dx)}{v^2}$$

then

$$\frac{dK}{ds} = \frac{(1 - s)(2s + 4) - (s^2 + 4s + 9)(-1)}{(1 - s)^2}$$

$$\frac{dK}{ds} = \frac{-s^2 + 2s + 13}{s^2 - 2s + 1}$$

Equating this to zero means that we must have

$$-s^2 + 2s + 13 = 0$$

The roots of this equation are

$$\frac{-2 \pm \sqrt{(4 + 52)}}{-2} = 1 \pm 3.7$$

Only the value -2.7 is feasible. Thus there is a breakaway point at -2.7.

9 There is one pair of complex poles and no complex zeros. The angle of departure of the locus at $K = 0$ from a complex pole is determined using the argument equation [6] and letting s be a point on the root locus very close to the pole $2 + j3$. Since there is an angle β to the zero of 90° plus $\tan^{-1} 2/3$ or 123.7°, an angle α to the other pole of 90° and an angle to the pole $2 + j3$ of α_1, then

$$123.7° - (\alpha_1 + 90°) = \pm \text{odd multiple of } 180°$$

$$33.7° - \alpha_1 = \pm 180°$$

Hence $\alpha_1 = 213.7°$ or 146.3°.

10 Figure 9.22 shows the completed root locus diagram.

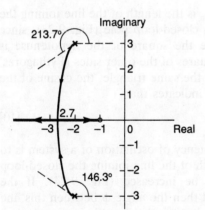

Fig. 9.22 Example 5

Interpretation of root locus diagrams

The root locus diagram shows the effect a variation in gain will have on the closed-loop characteristic equation roots and hence on the system dynamic behaviour. It enables the effect of modifying or adding poles and zeros to a system to be ascertained. Open-loop poles in such diagrams can be considered to act as 'sources' of loci and zeros as 'sinks' with K

increasing from zero at an open-loop pole to infinity at a zero.

The closed-loop transfer function of a second order system can be represented by

$$G(s) = \frac{\omega_n^2}{s^2 + 2\zeta\omega_n s + \omega_n^2} \qquad [7]$$

where ω_n is the natural angular frequency and ζ the damping ratio. If the damping ratio is between 0 and 1 then the poles are complex, the system producing an oscillating response. For this condition we can write

$$s^2 + 2\zeta\omega_n s + \omega_n^2 = (s + \sigma + j\omega)(s + \sigma - j\omega)$$
$$= s^2 + 2\sigma s + \sigma^2 + \omega^2$$

Thus

$$2\zeta\omega_n = 2\sigma$$

and hence

$$\zeta = \frac{\sigma}{\omega_n} \qquad [8]$$

and

$$\omega_n^2 = \sigma^2 + \omega^2 \qquad [9]$$

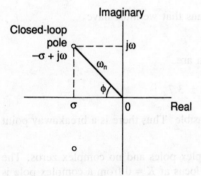

Fig. 9.23 Complex poles

Equation [9] means that ω_n is the length of the line joining the origin on the s-plane to a closed-loop pole (Fig. 9.23), since for a right-angled triangle the square of the hypotenuse is equal to the sum of the squares of the other sides (Pythagoras' theorem). Since σ/ζ is, for the same triangle, the cosine of the angle ϕ then equation [8] indicates that

$$\zeta = \cos\phi \qquad [10]$$

Thus if the angular frequency of oscillation of a system is to be increased then the length of the line joining the closed-loop pole to the origin must be increased (Fig. 9.24). If the damping is to be increased then the angle ϕ between this line and the real axis must be decreased since this increases $\cos\phi$ (Fig. 9.24).

As an illustration of the application of the above to a system, consider the root locus diagram in Fig. 9.25. For such a system the minimum value of the natural angular frequency will be when $K = 0$ since this gives the shortest length of line from a pole to the origin. This is also the minimum value of $\cos\phi$ and so the minimum value of the damping ratio. As K increases so ω_n and ζ increase until at the breakaway point $\zeta = 1$ and the damping is critical. Further increases in K result only in real roots and so there is no oscillation.

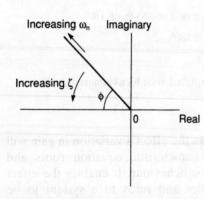

Fig. 9.24 Angular frequency and damping ratio

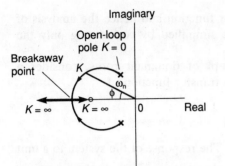

Fig. 9.25 Effect of changing K on ω_n and ζ

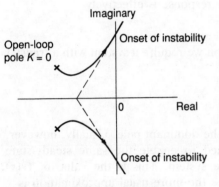

Fig. 9.26 Onset of instability

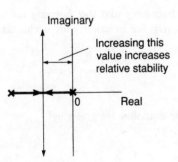

Fig. 9.27 Relative stability

Changing the natural angular frequency and/or the damping ratio results in changes in such parameters of system performance as the rise time, percentage overshoot and settling time (see Chapter 3). Thus, for example, the 2% settling time is given by (equation [36], Chapter 3)

$$t_s = \frac{4}{\zeta \omega_n}$$

Since equation [8] gives $\zeta = \sigma/\omega_n$, then

$$t_s = \frac{4}{\sigma} \qquad [11]$$

The rise time t_r of a system is given by (equation [29] Chapter 2)

$$t_r = \frac{\pi}{2\omega} \qquad [12]$$

Thus increasing ω decreases the rise time.

The root locus diagram can also be used to consider the effect of changing gain on the stability of the system. Thus for the root locus diagram given in Fig. 9.26 instability begins to occur when the value of K is such that the root locus diagram reaches the imaginary axis. The relative instability of a system can be judged by the proximity of its root loci to the imaginary axis, a system with root loci further away than those of another system is relatively more stable (Fig. 9.27).

The introduction of a zero into the left-hand side of the s-plane improves the relative stability of a system since it increases the angle of the asymptotes and so moves loci away from the imaginary axis. The angle of an asymptote is

$$\frac{\pi}{n - m}$$

where n is the number of poles and m the number of zeros. Increasing m, when $n > m$, brings the angle closer to $\pi/2$ and so brings the loci parallel to the imaginary axis. When loci are parallel to the axis then there is no value of K which will result in instability. The introduction of an extra pole has the opposite effect.

A higher-order system will be more complex than the systems discussed above. However if it has some poles or zeros closer to the origin than others then they will dominate the dynamic behaviour of the system and so the frequency and damping ratio for them alone will often be enough to describe the system behaviour. The roots nearest to the origin of the s-plane are called the *dominant roots*. There is reasonable dominance if the ratio of the real parts of the roots is greater than about 5. The transfer function and hence the analysis of the root loci can then be simplified by considering only the dominant roots.

To illustrate this concept of dominant roots, consider a system with a closed-loop transfer function of

$$G(s) = \frac{5}{(s + 1)(s + 5)}$$

The roots are -1 and -5. The response of the system to a unit impulse of 1 is

$$\theta_o(s) = \frac{5}{(s + 1)(s + 5)}$$

$$\theta_o = 5 \times 0.25(e^{-1t} - e^{-5t})$$

The term e^{-5t} can generally be neglected in comparison with the e^{-1t} term so that the response is effectively

$$\theta_o \approx 5 \times 0.25\,e^{-1t}$$

To obtain this as a solution we require a system with a transfer function of

$$G(s) = \frac{5 \times 0.25}{(s + 1)}$$

The $s = -1$ pole is thus the dominant pole. Usually, however, the numerator is modified to enable the same steady-state output to prevail for the system. This is the value of $G(s)$ when $s = 0$, i.e. 1. Hence the more usual approximation is

$$G(s) = \frac{1}{(s + 1)}$$

Example 6

What is the natural angular frequency and the damping ratio for a system having a closed-loop transfer function with a characteristic equation of

$$s^2 + 2s + 4$$

Answer

The roots of the characteristic equation are given by

$$\frac{-2 \pm \sqrt{(4 - 16)}}{2}$$

and are thus $-1 \pm j\sqrt{3}$. Hence the root $-1 + j\sqrt{3}$ forms a right-angled triangle with hypotenuse ω_n and sides of length 1 and $\sqrt{3}$ (Fig. 9.28), and so

$$\omega_n^2 = \sigma^2 + \omega^2$$

$$\omega_n^2 = 1^2 + 1.7^2 = 4$$

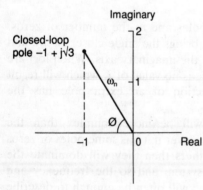

Fig. 9.28 Example 6

Hence $\omega_n = 2\,\text{rad/s}$. The damping ratio is the cosine of angle ϕ, and so

$$\zeta = \cos\phi = \tfrac{1}{2}$$

Example 7

A system has an open-loop transfer function of

$$G_o(s) = \frac{K}{(s + 2 + j2)(s + 2 - 2j)}$$

(a) Sketch the root locus diagram.

(b) What is (i) the angular frequency, (ii) the damping factor, (iii) the rise time and (iv) the 2% settling time for the system when $K = 10$?

(c) The system gives a steady-state error which it is proposed to eliminate by including in the forward path a block with a transfer function of $1/s$. What will now be the root locus diagram?

(d) What will be the modified 2% settling time when $K = 10$?

(e) How does the modification affect the relative stability for $K = 10$?

Answer

(a) Following the steps outlined earlier for the plotting of a root locus diagram:

   1 The system has unity feedback and thus the system transfer function is

   $$G(s) = \frac{K}{(s + 2 + j2)(s + 2 - j2) + K}$$

   The characteristic equation is thus

   $$(s + 2 + j2)(s + 2 - j2) + K = 0$$

   2 When $K = 0$ the characteristic equation becomes

   $$(s + 2 + j2)(s + 2 - j2) = 0$$

   The open-loop poles are thus $2 \pm j2$. There are no zeros.

   3 The equation is of second order and so there are two loci.

   4 There are no portions of the root loci on the real axis.

   5 The angles of the asymptotes will be, since $n - m = 2$, $\pi/2$ or $90°$.

   6 The intersection of the asymptotes with the real axis will be

   $$\frac{(-2 + j2) + (-2s - j2)}{2} = -2$$

   7 Any points of intersection of loci with the imaginary axes can be determined by putting $s = j\omega$ in the characteristic equation. This gives

   $$(j\omega + 2 + j2)(j\omega + 2 - j2) + K = 0$$

   $$-\omega^2 + j4\omega + 8 + K = 0$$

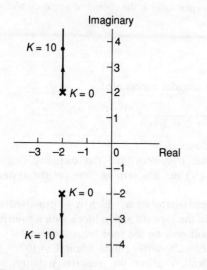

Fig. 9.29 Example 7

Hence, equating imaginary parts gives $\omega = 0$ and equating real parts $K = -8$. Thus the loci do not cross the imaginary axis, since for the loci we are concerned with only positive values of K occur.

8 Since the loci do not touch the real axis there is no breakaway point.

9 There is just one pair of complex poles and no zeros. The angle of departure of the locus at $K = 0$ from a complex pole is determined using the argument equation [6] and letting s be a point on the root locus very close to $(2 + j2)$. Since $\alpha_1 = 90°$ then the angle to the pole $(2 + j2)$ of α_2 is given by

$$(\alpha_2 + 90°) = \pm \text{odd multiple of } 180°$$

Hence $\alpha_2 = 90°$.

10 Figure 9.29 shows the completed root locus diagram.

(b) When $K = 10$ the characteristic equation becomes

$$(s + 2 + j2)(s + 2 - j2) + 10 = 0$$

which becomes

$$s^2 + 4s + 18 = 0$$

The roots are thus

$$\frac{-4 \pm \sqrt{(16 - 72)}}{2} = -2 \pm j\sqrt{14} = -2 \pm j3.7$$

(i) The angular frequency can be obtained using equation [9]

$$\omega_n^2 = \sigma^2 + \omega^2$$

$$\omega_n^2 = 2^2 + (\sqrt{14})^2$$

Hence $\omega_n = 4.2 \, \text{rad/s}$.

(ii) The damping ratio can be obtained using equation [10]

$$\zeta = \cos \phi = \frac{2}{4.2} = 0.48$$

(iii) The rise time is given by equation [12] as

$$t_r = \frac{\pi}{2\omega} = \frac{\pi}{2\sqrt{14}} = 0.42 \, \text{s}$$

(iv) The 2% settling time can be obtained using equation [11]

$$t_s = \frac{4}{\sigma} = 2\sigma$$

(c) The root loci for the modified transfer function can be obtained following the steps outlined earlier:

1 The system has unity feedback and thus the system transfer function is

$$G(s) = \frac{K}{s(s + 2 + j2)(s + 2 - j2) + K}$$

The characteristic equation is thus

$$s(s + 2 + j2)(s + 2 - j2) + K = 0$$

2   When $K = 0$ the characteristic equation becomes

$$s(s + 2 + j2)(s + 2 - j2) = 0$$

The open-loop poles are thus 0, and $2 \pm j2$. There are no zeros.

3   The equation is of third order and so there are three loci.

4   The portions of the root loci on the real axis are from 0 to infinity.

5   The angles of the asymptotes will be (since $n - m = 3$) $\pi/3$ or 60°, $3\pi/3$ or 180° and $5\pi/3$ or 300°.

6   The point of intersection of the asymptotes is

$$\frac{0 + (-2 + j2) + (-2 - j2)}{3} = -\frac{4}{3}$$

7   Any points of intersection of loci with the imaginary axes can be determined by putting $s = j\omega$ in the characteristic equation. This gives

$$j\omega(j\omega + 2 + j2)(j\omega + 2 - j2) + K = 0$$

$$-j\omega^3 - 4\omega^2 + j8\omega + K = 0$$

Hence, equating imaginary parts gives $-\omega^3 + 8\omega = 0$ and equating real parts $-4\omega^2 + K = 0$ and so $\omega = \pm\ 2.8$ and $K = 32$.

8   The breakaway point is determined from the characteristic equation, which is

$$s(s + 2 + j2)(s + 2 - j2) + K = 0$$

This can be simplified to

$$s^3 + 4s^2 + 8s + K = 0$$

$$K = -s^3 - 4s^2 - 8s$$

then

$$\frac{dK}{ds} = -3s^2 - 8s - 8$$

Equating this to zero means that we must have

$$3s^2 + 8s + 8 = 0$$

The roots of this equation are

$$\frac{-8 \pm \sqrt{(64 - 96)}}{6} = -1.3 \pm j0.9$$

There is thus no breakaway point on the real axis.

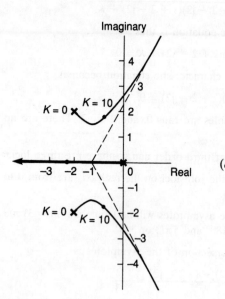

Fig. 9.30 Example 7

9 The angle of departure of the locus at $K = 0$ from a complex pole is determined using the argument equation [6] and letting s be a point on the root locus very close to the pole $2 + j2$.

$$-(\alpha_1 + 135° + 90°) = \pm \text{odd multiple of } 180°$$

$$-225° - \alpha_1 = \pm 180° \text{ or } \pm 540°$$

Hence $\alpha_1 = 45°$ or $315°$.

10 Figure 9.30 shows the completed root locus diagram.

(d) When $K = 10$ the characteristic equation becomes

$$s(s + 2 + j2)(s + 2 - j2) + 10 = 0$$

which becomes

$$s^3 + 4s^2 + 8s + 10 = 0$$

Since these roots will be on the loci in Fig. 9.30, one of the roots of the equation must be on the real axis and the other pair must be complex. The roots are -2.4, and $-0.8 \pm 1.9j$.

The 2% settling time can be obtained using equation [11]

$$t_s = \frac{4}{\sigma} = \frac{5}{0.8} = 6.25 \text{ s}$$

Thus introducing the $1/s$ element has increased the settling time.

(e) With $K = 10$ the original system gave $\sigma = -2$, with the modification it is -0.8. Thus the relative stability has been reduced. One way to overcome this and also reduce the settling time would be to reduce the value of K.

Example 8

A closed-loop system has three poles and no zeros. (a) What can be said about its stability and (b) how its relative stability can be improved?

Answer

(a) The angles of the asymptotes will be $\pi/3$, $3\pi/3$, and $5\pi/3$ and thus will intercept at some value of gain with the imaginary axis. Thus at some value of gain the system will become unstable.

(b) The relative stability can be improved by adding a zero. This will then give asymptotes at angles of $\pi/2$ and $3\pi/2$. None of these will give intercepts with the imaginary axis, but loci parallel to that axis, and thus the system has had its stability improved.

Example 9

A system has an open-loop transfer function of

$$G_o(s) = \frac{K}{s(s + 6)}$$

What is the gain when there is (a) critical damping, (b) a damping ratio of 0.6?

Answer

The closed-loop transfer function of the system will be

$$G(s) = \frac{K}{s(s+6)+K} = \frac{K}{s^2+6s+K}$$

Hence the roots are

$$\frac{-6 \pm \sqrt{(36-4K)}}{2} = -3 \pm j\sqrt{(K-9)}$$

(*a*) Critical damping occurs when the imaginary part of the root is zero, i.e. when $K - 9 = 0$. Thus $K = 9$.

(*b*) The damping factor is given by equation [8] as

$$\zeta = \frac{\sigma}{\omega_n}$$

This system has σ with the value 3 and thus for $\zeta = 0.6$ we have $\omega_n = 5$ rad/s. But equation [10] gives

$$\omega_n^2 = \sigma^2 + \omega^2$$

and so

$$\omega^2 = 5^2 - 3^2$$

and hence $\omega = 4$ rad/s. This is the value of the imaginary part of the root. Thus

$$4 = \sqrt{(K-9)}$$

and so $K = 25$. Figure 9.31 shows the root locus diagram.

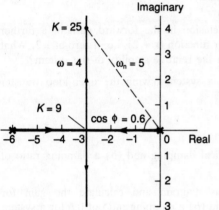

Fig. 9.31 Example 8

Problems

1 Sketch the root loci diagrams for the systems shown in Fig. 9.32 and for second order systems state the value of the gain K at which the systems are critically damped.

2 Sketch the root loci diagrams for systems having the following open-loop transfer functions, identifying poles, zeros, asymptotes, and breakaway points.

(*a*) $\dfrac{K(s+2)}{(s+1)(s+3)(s+4)}$

(*b*) $\dfrac{K(s+5)}{(s+2)(s+4+j3)(s+4-j3)}$

(*c*) $\dfrac{K}{(s+1)(s^2+9s+25)}$

(*d*) $\dfrac{K}{s^2+s+4}$

3 What are the natural angular frequencies and damping ratios for systems having closed-loop transfer functions with characteristic equations of:

(*a*) $s^2 + 4s + 16$.

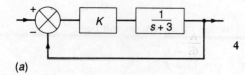

(a)

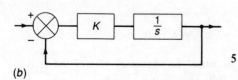

(b)

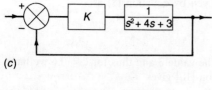

(c)

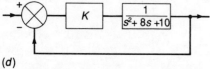

(d)

Fig. 9.32 Problem 1

(b) $s^2 + 6s + 12$.
(c) $s^2 + 2s + 10$.

4   What are the 2% settling times and rise times for a system having an open-loop transfer function of

$$\frac{K}{s(s + 2)}$$

when (a) $K = 4$, (b) $K = 16$?

5   A system giving the open-loop transfer function

$$\frac{K}{(s + 1)(s^2 + 9s + 25)}$$

is modified by the inclusion in the forward path of a further element with a transfer function $(s + 2)$, i.e. a zero at -2. What is the effect of this on the relative stability of the system?

6   What is the gain for a system having the open-loop transfer function of

$$G_o(s) = \frac{K}{s(s + 3)}$$

when there is (a) critical damping and (b) a damping ratio of 0.3?

7   Sketch the root locus diagram and calculate the gain for (a) critical damping and (b) a damping ratio of 0.6 for a system having an open-loop transfer function of

$$G_o(s) = \frac{K}{s(s + 1)}$$

10 Controllers

Introduction

This chapter is concerned with the selection of the appropriate form of controller for plant in a closed-loop control system and the determination of suitable parameters for that controller. The controller is the element in the closed-loop control system which has an input of the error signal and produces an output which becomes the input to the corrective element (see Chapter 1). The relationship between the output and the input to the controller is often called the *control law*. There are three forms of such law, *proportional*, *integral* and *derivative*, and they are considered in this chapter. In some systems it is necessary to enhance the performance of the controller and this is achieved by introducing additional elements, called *compensators*, into the control system. This alteration of performance is called *compensation*.

Proportional control

With proportional control, the output of the controller is directly proportional to its input, the input being the error signal e which is a function of time. Thus

$$\text{Output} = K_p e \qquad [1]$$

where K_p is a constant called the *proportional gain*. The output from the controller depends only on the size of the error at the instant of time concerned. The transfer function $G_c(s)$ for the controller is thus

$$G_c(s) = K_p \qquad [2]$$

The controller is effectively just an amplifier with a constant gain. A large error at some time produces a large controller output at that time. The constant gain, however, tends to exist only over a certain range of errors, this range being called the *proportional band*. A graph of output against error would be a straight line with a slope of K_p within the proportional band (Fig. 10.1).

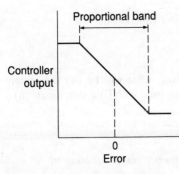

Fig. 10.1 Proportional control

It is usual to express controller output as a percentage of the total controller output possible. Thus a 100% change in controller output corresponds to an error change from one extreme of the proportional band to the other. Thus

$$K_{\text{p}} = \frac{100}{\text{proportional band}} \qquad [3]$$

Because the output is proportional to the input, if the input to the controller is an error in the form of a step then the output is also a step, being an exact scaled version of the input (Fig. 10.2). This is provided the controller is operating within its proportional band.

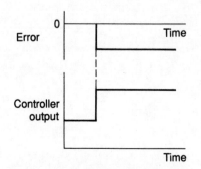

Fig. 10.2 System with proportional control

Proportional control is simple to apply, essentially just requiring some form of amplifier. This could be an electronic amplifier or a mechanical amplifier in the form of a lever (see later in this chapter), the control system with proportional control being of the form shown in Fig. 10.3. The result is in an open-loop transfer function of

$$G_{\text{o}}(s) = K_{\text{p}}G_{\text{p}}(s) \qquad [4]$$

where $G_{\text{p}}(s)$ is the transfer function of the plant.

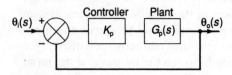

Fig. 10.3 System with proportional control

The main disadvantage of the system is that the controller does not introduce a $1/s$ or integrator in the forward path. This means that if the system was type 0 then the controller would not change this and it would remain type 0 with consequently steady-state errors (see Chapter 7). The controller does not introduce any new zeros or poles to the system, only determining the location of closed-loop poles. This is because the closed-loop transfer function with the controller, and unity feedback, is

$$G(s) = \frac{K_{\text{p}}G_{\text{p}}(s)}{1 + K_{\text{p}}G_{\text{p}}(s)}$$

and so the characteristic equation of $(1 + K_{\text{p}}G_{\text{p}}(s))$ has its values of roots affected by the value of K_{p}.

Example 1

If the plant in Fig. 10.3 has a transfer function of

$$G_{\text{p}}(s) = \frac{1}{s(s + 1)}$$

and is used with proportional control, what will be (a) the system type, (b) the steady-state errors when used with (i) a step input, (ii) a ramp input?

Answer

(a) The system will have an open-loop transfer function of

$$G_o(s) = \frac{K_p}{s(s + 1)}$$

Thus the system is type 1. See Chapter 7 for further discussion.

(b) (i) The steady-state error e_{ss} is given by (see Chapter 7)

$$e_{ss} = \lim_{s \to 0} \left[s \frac{1}{1 + G_o(s)} \theta_i(s) \right]$$

where, for a step input $\theta_i(s) = 1/s$. Thus

$$e_{ss} = \lim_{s \to 0} \left[s \frac{1}{1 + [K_p/s(s + 1)]} \frac{1}{s} \right]$$

$$= \frac{1}{\infty} = 0$$

Thus for a step input a type 1 system has zero steady-state error.

(ii) For a ramp input the input is $1/s^2$ and the steady-state error is

$$e_{ss} = \lim_{s \to 0} \left\{ s \frac{1}{1 + [K_p/s(s + 1)]} \frac{1}{s^2} \right\}$$

$$= \frac{1}{K_p}$$

Integral control

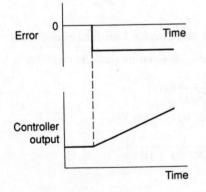

Fig. 10.4 Integral control

With integral control the output of the controller is proportional to the integral of the error signal e with time, i.e.

$$\text{Output} = K_i \int_0^t e \, dt \qquad [5]$$

where K_i is a constant called the *integral gain*. It has the unit of s^{-1}. Figure 10.4 shows what happens when the error is in the form of a step. The integral between t and 0 is in fact the area under the error graph between t and 0. Thus, since after the start of the error the area increases at a regular rate then the output from the controller must increase at a regular rate. The output at any time is thus proportional to the accumulation of the effects of past errors.

Taking the Laplace transform of equation [5] results in a transfer function, for the integral controller, of

$$G_c(s) = \frac{\text{output } (s)}{e(s)} = \frac{K_i}{s} \qquad [6]$$

Thus, for a system of the form shown in Fig. 10.5, integral control gives a forward-path transfer function of $(K_i/s)G_p(s)$ and hence an open-loop transfer function of

$$G_o(s) = \left(\frac{K_i}{s} \right) G_p(s) \qquad [7]$$

An advantage of integral control is that the introduction of an s term in the denominator increases the type number of the system by 1. Thus if the system had been type 0 the steady state-error that would have occurred with the step input disappears when integral control is present. A disadvantages of integral control is that a $(s - 0)$ term in the denominator means a pole has been introduced at the origin. Since no zeros are introduced the difference between the number of poles n and zeros m has been increased by 1. A consequence of this is that asymptote angles of root loci are decreased, i.e. they point more towards the right half of the s-plane, and thus the relative stability has been reduced.

$$\text{Asymptote angles} = \pm \frac{\pi}{n - m}, \frac{3\pi}{n - m}, \text{etc.}$$

Example 2

If the plant in Fig. 10.5 has a transfer function of

$$G_p(s) = \frac{1}{s(s + 1)}$$

and is used with integral control, what will be (a) the system type, (b) the steady-state errors when used with (i) a step input, (ii) a ramp input and (c) how does the stability compare with that which would occur if there was proportional control (as in Example 1)?

Answer

(a) The system will have an open-loop transfer function of

$$G_o(s) = \frac{K_i}{s^2(s + 1)}$$

Thus the system is type 2. See Chapter 7 for further discussion.

(b) (i) The steady-state error e_{ss} is given by (see Chapter 7)

$$e_{ss} = \lim_{s \to 0} \left[s \frac{1}{1 + G_o(s)} \theta_i(s) \right]$$

where, for a step input $\theta_i(s) = 1/s$. Thus

$$e_{ss} = \lim_{s \to 0} \left\{ s \frac{1}{1 + [K_i/s^2(s + 1)]} \frac{1}{s} \right\}$$

$$= \frac{1}{\infty} = 0$$

Thus for a step input a type 1 system has zero steady-state error.

(ii) For a ramp input the input is $1/s^2$ and the steady-state error is

$$e_{ss} = \lim_{s \to 0} \left\{ s \frac{1}{1 + [K_i/s^2(s + 1)]} \frac{1}{s^2} \right\}$$

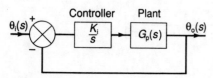

Controller   Plant

$\theta_i(s)$ $\frac{K_i}{s}$ $G_p(s)$ $\theta_o(s)$

Fig. 10.5 Integral control

$$= \frac{1}{\infty} = 0$$

Thus for a ramp input there is zero steady-state error, an improvement on the situation existing in Example 1 when there was proportional control.

(c) For the proportional control situation shown in Example 1 the system has a transfer function of

$$G(s) = \frac{K_p/[s(s + 1)]}{1 + K_p/s(s + 1)} = \frac{K_p}{s(s + 1) + K_p}$$

The characteristic equation is thus

$$s^2 + s + K_p = 0$$

The Routh array (see Chapter 8) for this is

$$\begin{array}{c|cc} s^2 & 1 & K_p \\ s^1 & 1 & \\ s^0 & K_p & \end{array}$$

The first column is all positive if K_p is greater than 0.

For the integral control situation the system has a transfer function of

$$G(s) = \frac{K_i/s^2(s + 1)}{1 + K_i/s^2(s + 1)} = \frac{K_i}{s^2(s + 1) + K_i}$$

The characteristic equation is thus

$$s^3 + s^2 + K_i$$

Since the equation lacks a term in s the system is unstable for all values of K_i. The Routh array would be

$$\begin{array}{c|cc} s^3 & 1 & 0 \\ s^2 & 1 & K_i \\ s^1 & -K_i & \\ s^0 & K_i & \end{array}$$

Thus a change from proportional control to integral control has, in this instance, resulted in instability. Figure 10.6 shows the root locus diagrams for the two situations, diagram (a) being with proportional control and (b) with integral control.

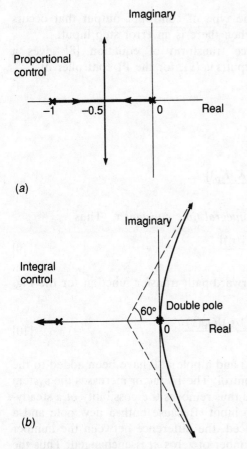

Proportional control

Integral control

(a)

(b)

Fig. 10.6 Example 2

Proportional plus integral control

The reduction in relative stability resulting from using integral control can be overcome, to some extent, by using proportional plus integral (PI) control (Fig. 10.7). For such a combination the controller output is

$$\text{Output} = K_p e + K_i \int_0^t e\, dt \qquad [8]$$

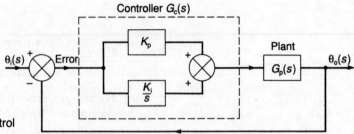

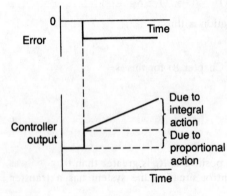

Fig. 10.7 Proportional plus integral control

Fig. 10.8 Proportional plus integral control

Figure 10.8 shows the type of controller output that occurs with such a system when there is an error step input.

Taking the Laplace transform of equation [8] gives a transfer function, output$(s)/e(s)$, for the PI controller of

$$G_c(s) = K_p + \frac{K_i}{s}$$

$$= \frac{sK_p + K_i}{s}$$

$$= \frac{K_p[s + (K_i/K_p)]}{s}$$

(K_p/K_i) is called the *integral time constant* τ_i. Thus

$$G_c(s) = \frac{K_p[s + (1/\tau_i)]}{s} \tag{9}$$

Consequently, the forward-path transfer function for the Fig. 10.7 system is

$$G_o(s) = \frac{K_p[s + (1/\tau_i)]G_p(s)}{s} \tag{10}$$

Thus a zero of $-(1/\tau_i)$ and a pole at 0 have been added to the system by using PI control. The $1/s$ factor increases the system type number by 1 and thus removes the possibility of a steady-state error for a step input. Because both a new pole and a new zero are introduced, the difference between the number of poles n and the number of zeros m is unchanged. Thus the asymptote angles for the root loci are unchanged.

$$\text{Asymptote angles} = \pm \frac{\pi}{n-m}, \quad \frac{3\pi}{n-m}, \quad \text{etc.}$$

However, the point of intersection of the asymptotes with the real axis is moved nearer to the origin and consequently there is some reduction in relative stability.

$$\text{Intersection point} = \frac{\text{sum of poles} - \text{sum of zeros}}{n-m}$$

Adding a pole at $s = 0$ and a zero at $s = -(1/\tau_i)$ results in the intersection point changing by $+(1/\tau_i)/(n - m)$ and so becoming more positive and moving to the right and so nearer to the origin. The reduction in relative stability is however not so much as with integral control alone.

The position of the introduced zero is determined by the value of the integral gain K_i, i.e. this is the same as saying it is determined by integral time constant τ_i. The proportional gain K_p determines the closed-loop pole positions (see the section on proportional control earlier in this chapter).

Example 3

If the plant in Fig. 10.7 has a transfer function of

$$G_p(s) = \frac{1}{s(s + 1)}$$

and is used with proportional plus integral control, what will be (a) the system type, (b) the steady-state errors when used with (i) a step input, (ii) a ramp input and (c) how does the stability compare with that which would occur if there was (i) proportional control (as in Example 1), (ii) integral control (as in Example 2)? The integral time constant is 2 s.

Answer

(a) The system will have an open-loop transfer function given by equation [10] as

$$G_o(s) = \frac{K_p[s + (1/\tau_i)]G_p(s)}{s} = \frac{K_p(s + 0.5)}{s^2(s + 1)}$$

Thus the system is type 2. See Chapter 7 for further discussion.

(b) (i) The steady-state error e_{ss} with a step input is for a type 2 system zero (see Example 2)

    (ii) For a ramp input a type 2 system gives a zero steady-state error (see Example 2). The system is thus better than the proportional control alone and the same as the integral control alone.

(c) For the proportional plus integral control situation the system has a transfer function of

$$G(s) = \frac{K_p[s + (1/\tau_i)]G_p(s)/s}{1 + K_p[s + (1/\tau_i)]G_p(s)/s}$$

$$= \frac{K_p[s + 0.5][1/s(s + 1)]/s}{1 + K_p[s + 0.5][1/s(s + 1)]/s}$$

$$= \frac{K_p[s + 0.5]}{s^2(s + 1) + K_p(s + 0.5)}$$

The characteristic equation is thus

$$s^3 + s^2 + K_ps + 0.5K_p = 0$$

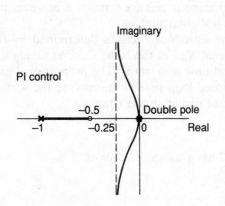

Fig. 10.9 Example 3

Derivative control

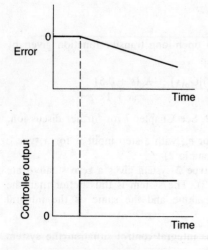

Fig. 10.10 Derivative control

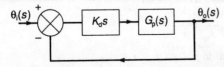

Fig. 10.11 Derivative control

The Routh array (see Chapter 8) for this is

$$
\begin{array}{c|cc}
s^3 & 1 & K_p \\
s^2 & 1 & 0.5K_p \\
s^1 & 0.5K_p \\
s^0 & 0.5K_p
\end{array}
$$

The first column is all positive if $0.5K_p$ is greater than 0. The system is thus stable when K_p is greater than 0. The adding of proportional control to integral control has resulted in restoring stability. Figure 10.9 shows the root locus diagram for the system. This should be compared with the root locus diagrams in Fig. 10.6 for proportional and integral control when separate. Compared with only proportional control the relative stability is reduced, the asymptotes being nearer to the imaginary axis with PI control, but compared with integral control alone the PI control has pushed the root loci into the left-hand side of the s-plane.

With the derivative form of controller, the controller output is proportional to the rate of change of the error e with time, i.e.

$$
\text{Output} = K_d \frac{de}{dt} \qquad [11]
$$

where K_d is the *derivative gain* and has the units of s. Figure 10.10 shows what happens when there is a ramp error input. With derivative control, as soon as the error signal begins there can be quite a large controller output since the output is proportional to the rate of change of the error signal and not its value. It can thus provide a large corrective action before a large error actually occurs. However if the error is constant then there is no corrective action, even if the error is large. Derivative control is thus insensitive to constant or slowly varying error signals and is consequently not used alone but combined with other forms of controller.

Taking the Laplace transform of equation [11] gives, for derivative control, a transfer function output$(s)/e(s)$ of

$$
G_c(s) = K_d s \qquad [12]
$$

Hence for the closed-loop system shown in Fig. 10.11, the presence of the derivative controller results in an open-loop transfer function of

$$
G_o(s) = \frac{K_d s G_p(s)}{1 + K_d s G_p(s)} \qquad [13]
$$

If the plant is a type 1 or higher system then the application of derivative action is to cancel out an s in the denominator and so reduce the order by 1. However, as mentioned earlier, derivative action is not used alone but only in conjunction with other forms of controller. Where it is used it speeds up the response of a system to errors.

There are difficulties in implementing a derivative control law so in practice an approximate derivative control is obtained by using a lead compensator (see later in this chapter). This has a transfer function of the form $K(s + z)/(s + p)$, with $p > z$.

Proportional plus derivative control

If derivative control is used with proportional control (Fig. 10.12) then the open-loop transfer function becomes

$$G_o(s) = (K_p + K_d s)G_p(s)$$

$$G_o(s) = K_d[(1/\tau_d) + s]G_p(s) \qquad [14]$$

where $\tau_d = K_p/K_d$ and is called the *derivative time constant*. With this form of control a zero at $s = -1/\tau_d$ has been introduced. Also there has been no change in the system type and hence steady-state errors.

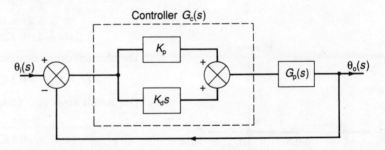

Fig. 10.12 Proportional plus derivative control

Example 4

If the plant in Fig. 10.12 has a transfer function of

$$G_p(s) = \frac{1}{s(s + 1)}$$

and is used with proportional plus derivative control, what will be (*a*) the system type, (*b*) the steady-state errors when used with (i) a step input, (ii) a ramp input and (*c*) what is the condition for stability? The derivative time constant is 2 s.

Answer

(*a*) The system will have an open-loop transfer function given by equation [14] as

$$G_o(s) = K_d[(1/\tau_d) + s]G_p(s) = \frac{K_d(s + 0.5)}{s(s + 1)}$$

Thus the system is type 1. See Chapter 7 for further discussion.

(b) (i) The steady-state error e_{ss} is zero for a step input with a type 1 system. See Chapter 7 and also Example 1 in this chapter.

(ii) For a ramp input the input $\theta_i(s)$ is $1/s^2$ and the steady-state error is

$$e_{ss} = \lim_{s \to 0} \left[s \frac{1}{1 + G_o(s)} \theta_i(s) \right]$$

$$e_{ss} = \lim_{s \to 0} \left\{ s \frac{1}{1 + [K_d(s + 0.5)/s(s + 1)]} \frac{1}{s^2} \right\}$$

$$= \frac{1}{0.5 K_d}$$

Since $(1/\tau_d) = 0.5 = K_p/K_d$ then

$$e_{ss} = \frac{1}{K_p}$$

(c) For the proportional plus differential control situation the system has a transfer function of

$$G(s) = \frac{K_d(s + 0.5)G_p(s)}{1 + K_d(s + 0.5)G_p(s)}$$

$$= \frac{K_d(s + 0.5)[1/s(s + 1)]}{1 + K_d[s + 0.5][1/s(s + 1)]}$$

$$= \frac{K_d(s + 0.5)}{s(s + 1) + K_d(s + 0.5)}$$

The characteristic equation is thus

$$s^2 + (1 + K_d)s + 0.5K_d = 0$$

The Routh array (see Chapter 8) for this is

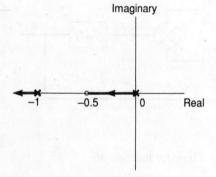

Imaginary

−1   −0.5   0   Real

Fig. 10.13 Example 4

$$\begin{array}{c|cc} s^2 & 1 & 0.5K_d \\ s^1 & 1 + K_d & \\ s^0 & 0.5K_d & \end{array}$$

The first column is all positive and the system stable if K_d is positive. Figure 10.13 shows the root locus diagram for the system. This should be compared with the root locus diagrams in Fig. 10.6 for proportional and integral control when separate.

PID control

A proportional plus integral plus derivative (PID) controller, or so-called *three-term controller*, with a system of the form shown in Fig. 10.14 will give an output, for an input of an error e, of

$$\text{output} = K_p e + K_i \int_0^t e \, dt + K_d \frac{de}{dt} \tag{15}$$

The transfer function, output $(s)/e(s)$, of the controller is thus

$$G_c(s) = K_p + \frac{K_i}{s} + K_d s \tag{16}$$

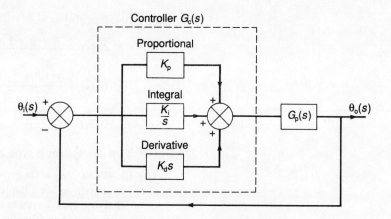

Fig. 10.14 PID control

Since the integral time constant τ_i is K_p/K_i and the derivative time constant τ_d is K_p/K_d equation [15] can be written as

$$G_c(s) = K_p\left(1 + \frac{K_i}{K_p s} + \frac{K_d s}{K_p}\right)$$

$$G_c(s) = K_p\left(1 + \frac{1}{\tau_i s} + \tau_d s\right) \qquad [17]$$

The open-loop transfer function for the system shown in Fig. 10.14 is

$$G_o(s) = G_c(s)G_p(s) = K_p\left(1 + \frac{1}{\tau_i s} + \tau_d s\right)G_p(s)$$

$$G_o(s) = \frac{K_p(\tau_i s + 1 + \tau_i \tau_d s^2)G_p(s)}{\tau_i s} \qquad [18]$$

Thus the PID controller has increased the number of zeros by 2 and the number of poles by 1. Also the $1/s$ factor increases the type number by 1. The above equation has assumed that an ideal differentor has been employed. In practice, as indicated earlier in this chapter, a lead compensator is used.

Example 5

If the plant in Fig. 10.14 has a transfer function of

$$G_p(s) = \frac{1}{s(s + 1)}$$

and is used PID control, what will be (a) the system type, (b) the steady-state errors when used with (i) a step input, (ii) a ramp input, (c) the positions of the open-loop zeros and poles, and (d) the condition for stability? The derivative time constant is $0.5\,\text{s}$ and the integral time constant $2\,\text{s}$.

Answer

(a) The system will have an open-loop transfer function given by equation [18] as

$$G_o(s) = \frac{K_p(\tau_i s + 1 + \tau_i \tau_d s^2) G_p(s)}{\tau_i s}$$

$$G_o(s) = \frac{K_p(s^2 + 2s + 1) G_p(s)}{2s} = \frac{K_p(s^2 + 2s + 1)}{2s^2(s + 1)}$$

$$G_o(s) = \frac{K_p(s + 1)}{2s^2}$$

Thus the system is type 2. See Chapter 7 for further discussion.

(b) (i) The steady-state error e_{ss} is zero for a step input with a type 2 system. See Chapter 7 and also Example 2 in this chapter.

(ii) For a type 2 system with a ramp input the steady-state error is zero. See Chapter 7 and also Example 2 in this chapter.

(c) The open-loop transfer function in part (a) above indicates that the system has an open-loop zero of –1 and poles of 0 and 0. One of the original poles has been cancelled by a zero introduced by the controller.

(d) For the PID control situation the system has a transfer function of

$$G(s) = \frac{K_p(s + 1)/2s^2}{1 + K_p(s + 1)/2s^2}$$

$$= \frac{K_p(s + 1)}{2s^2 + K_p(s + 1)}$$

The characteristic equation is thus

$$2s^2 + K_p s + K_p = 0$$

The Routh array (see Chapter 8) for this is

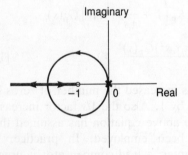

Imaginary

−1   0   Real

Fig. 10.15 Example 5

$$\begin{array}{c|cc} s^2 & 2 & K_p \\ s^1 & K_p & \\ s^0 & K_p & \end{array}$$

The first column is all positive and the system stable if K_p is positive. Figure 10.15 shows the root locus diagram for the system.

Adjustment of controller gains

The use of proportional control requires just one variable to be selected, the proportional gain K_p, for the control system to have the required dynamic behaviour. The use of a PI controller requires the selection of two variables, the proportional gain K_p and the integral gain K_i. With a PID controller three variables have to be selected: the proportional gain K_p, the integral gain K_I and the derivative gain K_d. The selection of these variables enables the locations of the poles and zeros introduced by the controller to be determined and hence affect the stability of the control system.

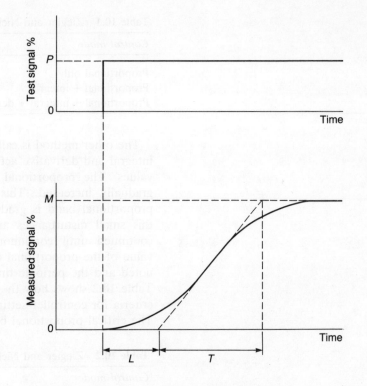

Fig. 10.16 Process reaction curve

The term *tuning* is used to describe the process of selecting the best controller settings. There are a number of methods of doing this: here just two methods will be discussed, both by Ziegler and Nichols. Both methods are based on experiments and analysis and are useful rules of thumb that are used frequently. The first method is often called the *process reaction curve method*. The procedure with this method is to open the control loop so that no control action occurs. Generally, the break is made between the controller and the correction unit. A test input signal is then applied to the correction unit and the response of the measured process variable determined, i.e. the error signal. The test signal should be as small as possible. Figure 10.16 shows the form of test signal and a typical response. The graph of measured signal plotted against time is called the *process reaction curve*.

The test signal P is expressed as the percentage change in the correction unit. The measured variable is expressed as the percentage of the full-scale range. A tangent is drawn to give the maximum gradient of the graph. For Figure 10.16 the maximum gradient R is M/T. The time between when the test signal started and when this tangent intersects the graph time axis is termed the lag L. Table 10.1 gives the criteria recommended by Ziegler and Nichols for control settings based on the values of P, R and L.

Table 10.1 Ziegler and Nichols process reaction curve criteria

Control mode	K_p	K_i	K_d
Proportional only	P/RL		
Proportional + integral	$0.9P/RL$	$1/3.33L$	
Proportional + integral + derivative	$1.2P/RL$	$1/2L$	$0.5L$

The other method is called the *ultimate cycle method*. First, integral and derivative actions are reduced to their minimum values. The proportional constant K_p is set low and then gradually increased. This is the same as saying that the proportional band is gradually made narrower. While doing this small disturbances are applied to the system. This is continued until continuous oscillations occur. The critical value of the proportional constant K_{pc} at which this occurs is noted and the periodic time of the oscillations T_c measured. Table 10.2 shows how the Ziegler and Nichols recommended criteria for controller settings are related to this value of K_{pc}. The critical proportional band is $100/K_{pc}$.

Table 10.2 Ziegler and Nichols ultimate cycle criteria

Control mode	K_p	K_i	K_d
Proportional alone	$0.5K_{pc}$		
Proportional + integral	$0.45K_{pc}$	$1.2/T_c$	
Proportional + integral + derivative	$0.6K_{pc}$	$2.0/T_c$	$T_c/8$

Example 6

Determine the settings of K_p, K_i and K_d required for a three-mode controller which gave the process reaction curve shown in Fig. 10.17 when the test signal was a 6% change in the control valve position.

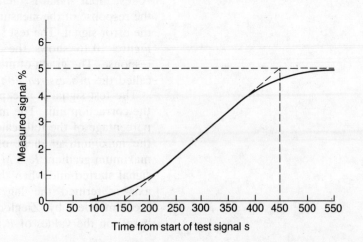

Fig. 10.17 Example 6

Answer

Drawing a tangent to the maximum gradient part of the graph gives a lag L of 150 s and a gradient R of $5/300 = 0.017$ %/s. Hence

$$K_p = \frac{1.2P}{RL} = \frac{1.2 \times 6}{0.017 \times 150} = 2.82$$

$$K_i = \frac{1}{1.2L} = \frac{1}{1.2 \times 150} = 0.0056 \, \text{s}^{-1}$$

$$K_d = 0.5L = 0.5 \times 150 = 75 \, \text{s}$$

Example 7

When tuning a three mode control system by the ultimate cycle method it was found that oscillations begin when the proportional band is decreased to 30%. The oscillations have a periodic time of 500 s. What are the suitable values of K_p, K_i and K_d?

Answer

The critical value of K_{pc} is 100/critical proportional band and so $100/30 = 3.33$. Thus, using the criteria given in Table 10.2,

$$K_p = 0.6K_{pc} = 0.6 \times 3.33 = 2.0$$

$$K_i = \frac{2}{T_c} = \frac{2}{500} = 0.004 \, \text{s}^{-1}$$

$$K_d = \frac{T_c}{8} = \frac{500}{8} = 62.5 \, \text{s}$$

Velocity feedback

In many systems involving the positioning of some object, e.g. a robot arm, a requirement is that the system should respond quickly to errors and not produce excessive oscillations or overshoots. This can be achieved by incorporating within the main feedback loop a minor feedback path which introduces what is called velocity feedback. The term *velocity feedback* is used to describe a feedback loop where the signal fed back is not the value of the output but the rate of change of the output with time. For such feedback the output of the feedback path is related to its input by

$$\text{Output} = K_v \frac{d\theta_o}{dt} \qquad [19]$$

and so the feedback path has a transfer function of

$$H(s) = K_v s \qquad [20]$$

K_v is a constant, the feedback gain, with the unit s.

The term *position feedback* is used in such circumstances for the output value feedback. These terms arise from the early use of control systems to control the position of some object,

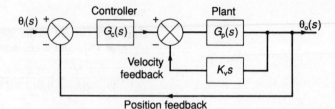

Fig. 10.18 System with velocity and position feedback

the position feedback then being a measure of the object's position with velocity feedback being a measure of the velocity of the object. Figure 10.18 shows a system with both forms of feedback. For such a system, the open-loop transfer function is

$$G_o(s) = \frac{G_c(s)G_p(s)}{1 + G_p(s)K_v s} \qquad [21]$$

and the closed-loop transfer function is

$$G(s) = \frac{G_c(s)G_p(s)}{1 + G_p(s)K_v s + G_p(s)G_c(s)} \qquad [22]$$

The effect of the velocity feedback has been to introduce a $G_p(s)K_v s$ term into the denominator and so the characteristic equation.

The effect of this can be seen by considering an example. Consider a system with proportional control, gain K_p, and a plant with transfer function $1/s(s + a)$. The open-loop transfer function is

$$G_o(s) = \frac{K_p/s(s + a)}{1 + K_v s/s(s + a)}$$

$$= \frac{K_p}{s(s + a) + K_v s}$$

$$= \frac{K_p}{s(s + a + K_v)} \qquad [23]$$

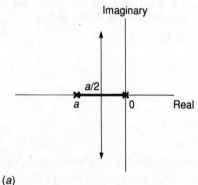

(a)

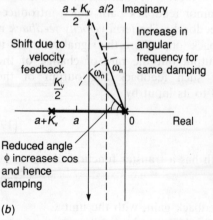

(b)

Fig. 10.19 The effect of velocity feedback: (a) without, and (b) with velocity feedback

The velocity feedback has modified the denominator, and hence the position of the open-loop roots. The roots, which were in the absence of velocity feedback 0 and a, are now 0 and $-(a + K_v)$. Figure 10.19 shows the effect on the root locus diagram. As a consequence of the velocity feedback the relative stability has been improved, the damping increased for the same natural angular frequency and the natural angular frequency increased for the same damping. The percentage overshoot is (Equation [33], Chapter 3)

$$\exp\left[\frac{-\zeta}{\sqrt{(1 - \zeta^2)}}\right] \times 100\%$$

Since $\zeta = \cos\phi$, then

$$\text{Overshoot} = \exp\left[\frac{-\cos\phi}{\sqrt{(1 - \cos^2\phi)}}\right] \times 100\%$$

$$= \exp\left(\frac{-\cos\phi}{\sqrt{\sin^2\phi}}\right) \times 100\%$$

Thus the percentage overshoot is

$$\text{Overshoot} = \exp(-1/\tan\phi) \times 100\% \qquad [24]$$

The effect of including velocity feedback is to reduce ϕ for a particular value of natural angular frequency. This means a reduction in $\tan\phi$ and consequently a decrease in the percentage overshoot.

Example 8

A closed-loop system has a proportional controller of gain K_p, a plant of transfer function

$$G_p = \frac{1}{s(s + 1)}$$

and unity position feedback. (a) What proportional gain is required for the natural angular frequency to be 2 rad/s? (b) What is the damping ratio at that frequency? (c) If velocity feedback was introduced, as in Fig. 10.18, what would the velocity gain need to be for the damping ratio to be doubled for the same angular frequency?

Answer

The root locus diagrams for the system without and then with velocity feedback will be of the same form as those given in Fig. 10.19.

(a) The open-loop transfer function of the system is

$$G_o(s) = \frac{K_p}{s(s + 1)}$$

The open-loop roots are those at 0 and -1. The closed-loop transfer function is

$$G(s) = \frac{K_p/s(s + 1)}{1 + K_p/s(s + 1)} = \frac{K_p}{s(s + 1) + K_p}$$

This has the characteristic equation

$$s^2 + s + K_p = 0$$

The roots of this equation are

$$\frac{-1 \pm \sqrt{(1 - 4K_p)}}{2} = -0.5 \pm j\sqrt{(K_p - 0.25)}$$

Thus the natural angular frequency ω_n is (Fig. 10.20)

$$\omega_n^2 = 0.5^2 + [\sqrt{(K_p - 0.25)}]^2$$

Hence for $\omega_n = 2$ rad/s then $K_p = 4.0$.

(b) The damping factor ζ is $\cos\phi$ and hence, using Fig. 10.20,

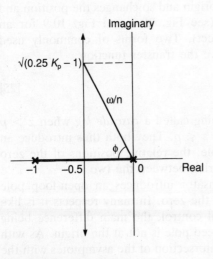

Fig. 10.20 Example 8

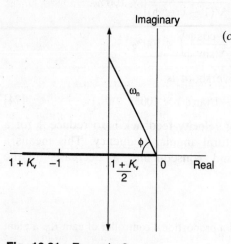

$$\zeta = \frac{0.5}{2} = 0.25$$

(c) With velocity feedback the open-loop transfer function becomes, as indicated by equation [23] above,

$$G_o(s) = \frac{K_p}{s(s + 1 + K_v)}$$

The open-loop poles thus becomes 0 and $-(1 + K_v)$. The root locus diagram then becomes as shown in Fig. 10.21. Then

$$\zeta = \cos \phi = \frac{\frac{1}{2}(1 + K_v)}{\omega_n}$$

Since ω_n is to remain unchanged at 2 rad/s and ζ is to be twice 0.25, then

$$\tfrac{1}{2}(1 + K_v) = 0.5 \times 2$$

and so $K_v = 1$.

Fig. 10.21 Example 8

Compensation

Compensators can be defined as components inserted into a control system to enhance the performance of the controller. If a controller is considered to be basically a proportional controller then, for example, a PI controller can be considered to be a proportional controller with an integral compensator. When the compensator is included in the forward path of the control loop then it is said to be a *cascade compensator*. Thus the integral compensator is a cascade integral compensator. The effect of including such a compensator has been discussed earlier in this chapter, there being mainly an improvement with respect to steady-state errors and a reduction in stability.

Compensators in improving performance are being used to alter and reshape the root locus. Thus the integral compensator introduces a pole at the origin and so changes the position and shape of the root loci (see Fig. 10.6 and Fig. 10.9 for an example of such an effect). Two forms of commonly used cascade compensator have the transfer function

$$G_c(s) = \frac{K(s + z)}{(s + p)} \tag{25}$$

with the compensator being called a *cascade lag* when $z > p$ and a *cascade lead* when $z < p$. They both thus introduce an open-loop zero and a pole, the relative positions of the zero and pole, however, differing between the two.

A cascade lag compensator introduces an open-loop pole closer to the origin than the zero. In many respects it is like proportional plus integral control, the main difference being however that the introduced pole is not at the origin. As with PI control, the point of intersection of the asymptotes with the real axis is moved to the right and so there is some reduction in stability. The angle of the asymptotes is unchanged since there is no change in $(n - m)$. However, because there is no $1/s$ factor introduced there is no change in the type number.

A cascade lead compensator introduces a zero closer to the origin than a pole. In many respects it is like proportional plus derivative control, the main difference being that the zero is not introduced at the origin. As with PD control, the point of intersection of the asymptotes with the real axis is moved to the left and so there is an improvement in stability. The angle of the asymptotes is unchanged since there is no change in $(n - m)$. No $1/s$ factor is introduced and so there is no change in the type number.

Example 9

Show the effect on the root locus diagram of a system with an open-loop transfer function

$$G_o(s) = \frac{K}{s(s + 1)}$$

of introducing into the forward path a lead compensator with transfer function

$$G_c(s) = \frac{s + 2}{s + 8}$$

and explain the effect on the relative stability.

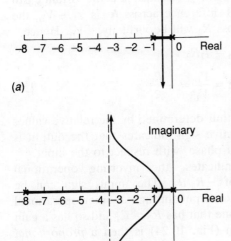

(a)

(b)

Fig. 10.22 Example 9

Answer

The uncompensated system has open-loop roots at 0 and -1 and no zeros. The asymptotes are at an angle of $\pi/2$ and intersect the real axis at $-1/2$, since $n - m = 2$. The locus follows the real axis between the two roots. The loci break away from the real axis at -0.5. The root locus diagram is consequently as shown in Fig. 10.22(*a*).

The compensated system has an open-loop transfer function of

$$G_o(s) = \frac{K(s + 2)}{s(s + 1)(s + 8)}$$

There are thus open-loop roots at 0, -1 and -8 with a zero at -2. The asymptotes are at an angle of $\pi/2$ and intersect the real axis at $-7/2$, since $n - m = 2$. The locus follows the real axis between 0 and -1, and -2 and -8. The breakaway point from the real axis is still about -0.5. The root locus diagram is thus as shown in Fig. 10.22(*b*).

Since the intercept of the asymptotes with the real axis has been moved from -0.5 to -3.5 there is an improvement in the relative stability.

Implementation of control laws

With electrical control systems operational amplifiers are often used as the basis for generating the required control laws. Figure 10.23 shows the basic form of such an amplifier when

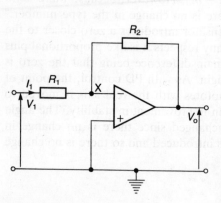

Fig. 10.23 The inverting form of operational amplifier

the connections made to it are for its use as an inverting amplifier. The amplifier has two inputs, referred to as the inverting input (−) and the non-inverting input (+). For use as an inverting amplifier the input is connected through a resistor R_1 to the inverting amplifier input and the amplifier non-inverting input is connected to earth. A feedback path is provided via resistor R_2. The operational amplifier has a very large transfer function, of the order of 100 000 or more, and the change in its output voltage is generally limited to about $\pm 10\,\text{V}$. To produce this output with this transfer function the input voltage must be between 0.001 V and −0.0001 V. This is virtually zero and so point X is at virtually earth potential. For this reason it is called a virtual earth. The potential difference across R_1 is $V_I - V_x$, hence the input potential V_I can be considered to be across R_1, and so

$$V_I = I_1 R_1$$

The operational amplifier has a very high impedance and so virtually no current flows through X into it. Hence the current through I_1 flows on through R_2. Because X is the virtual earth then, since the potential difference across R_2 is $V_x - V_o$, the potential difference across R_2 will be virtually $- V_o$. Hence

$$-V_o = I_1 R_2$$

$$\text{Transfer function} = \frac{V_o}{V_I} = -\frac{R_2}{R_1} \qquad [26]$$

The transfer function is thus determined by the relative values of R_2 and R_1. The negative sign indicates that the output is inverted, i.e. 180° out of phase, with respect to the input.

As equation [26] indicates, the inverting operational amplifier has a gain of $-R_2/R_1$. The minus sign can be eliminated by passing the output through another operational amplifier, but this time one that has $R_2 = R_1$ and so has a gain of −1. The combination (Fig. 10.24) is then a *proportional controller* with

$$K_p = \frac{R_2}{R_1} \qquad [27]$$

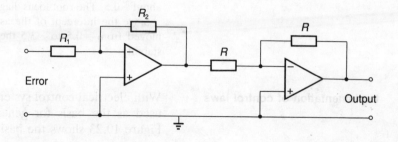

Fig. 10.24 Proportional controller

An integral controller can be produced if a capacitor replaces the feedback resistor (Fig. 10.25). For a capacitor, equation [20] Chapter 2 gives

$$v = \frac{1}{C} \int i \, dt$$

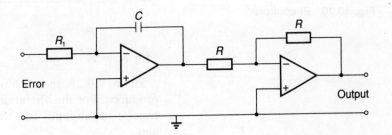

Fig. 10.25 Integral controller

Thus

$$V(s) = \frac{I(s)}{Cs}$$

and so the impedance Z is given by

$$Z(s) = \frac{1}{Cs} \qquad [28]$$

For the operational amplifier circuit with the capacitor feedback, the basic equation [26] can be written as

$$\text{Transfer function} = -\frac{Z_2}{Z_1} \qquad [29]$$

Since $Z_1(s) = R_1$ and $Z_2(s) = 1/Cs$, then

$$\text{Transfer function} = -\frac{1}{R_1 Cs}$$

This circuit when combined with another operational circuit of transfer function -1, as in Fig. 10.25, gives

$$\text{Transfer function} = \frac{(1/R_1 C)}{s} \qquad [30]$$

and hence is an integral controller with $K_i = (1/R_1 C)$. Figure 10.26 shows how the circuit can be adapted to give a PI controller. For this circuit

$$Z_2(s) = R_2 + \frac{1}{Cs}$$

and so

$$\text{Transfer function} = \frac{R_2 + 1/Cs}{R_1} = \frac{R_2}{R_1} + \frac{(1/R_1 C)}{s} \qquad [31]$$

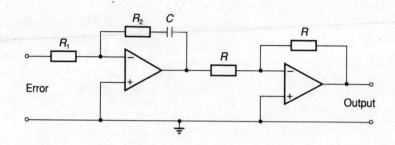

Fig. 10.26 PI controller

with $K_p = R_2/R_1$ and $K_i = 1/R_1C$.

Figure 10.27 shows how a derivative controller can be produced. For the operational amplifier with the capacitor and resistor in the input line we have, for equation [29], $Z_2 = R_2$ and

$$Z_1(s) = R_1 + \frac{1}{Cs} = \frac{R_1Cs + 1}{Cs}$$

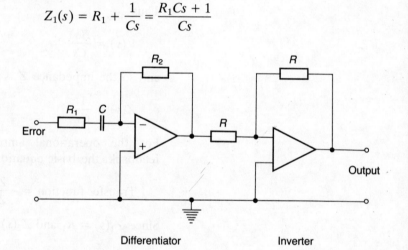

Fig. 10.27 Derivative controller

Thus equation [29] gives

$$\text{Transfer function} = -\frac{R_sCs}{R_1Cs + 1}$$

Combining this circuit with one having a transfer function of -1 gives

$$\text{Transfer function} = \frac{R_2Cs}{R_1Cs + 1} \tag{32}$$

The circuit has thus a derivative gain K_d of R_2C. Figure 10.28 shows how the circuit can be modified to give a PD controller.

Figure 10.29 shows how operational amplifier circuits of the PI controller and the D controller can be combined to give a PID controller. The transfer function is then

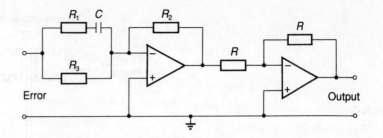

Fig. 10.28 PD controller

$$\text{Transfer function} = \frac{R_2}{R_1} + \frac{(1/R_1C_1)}{s} + \frac{R_4C_2s}{R_3C_2s + 1} \qquad [33]$$

with $K_p = R_2/R_1$, $K_i = 1/R_1C_1$ and $K_d = R_4C_2$.

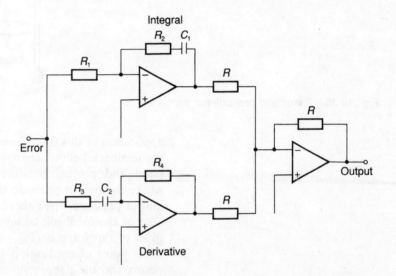

Fig. 10.29 PID controller

Pneumatic forms of proportional, integral and derivative controllers are used in many process control systems. Figure 10.30 shows the basic form of a pneumatic proportional controller. When the process pressure equals the set-point pressure the flapper-nozzle arrangement gives the output corresponding to zero error. When the process pressure changes from this value the flapper rotates and changes the gap between flapper and nozzle. The result is a change in the output pressure. This pressure changes until the feedback bellows exerts a force to balance that due to the process bellows.

The force due to the pressure difference between the set point and process bellows is $(P_p - P_s)A_1$, where P_p is the process pressure and P_s is the set point pressure. A_1 is the effective area of the bellows, both being assumed to be the same. The turning moment about the pivot of this force is $(P_p - P_s)A_1d_1$, where d_1 is the distance of the point of

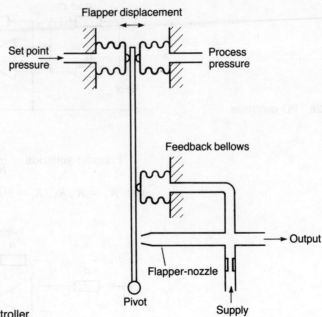

Fig. 10.30 Pneumatic proportional controller

application of this force from the pivot point. The force due to the feedback bellows changes from that occurring when the set point and process pressures were equal by $(P_{out} - P_o)A_2$, where the output pressure is P_{out} and P_o was the value of the output pressure when there was no error. A_2 is the effective area of the feedback bellows. The turning moment about the pivot of this force is $(P_{out} - P_o)A_2d_2$ where d_2 is the distance of its point of application from the pivot point. Equilibrium occurs and the flapper stops moving when

$$(P_{out} - P_o)A_2d_2 = (P_p - P_s)A_1D_1$$

and so

Change in output pressure $= P_{out} - P_o = K_p(P_p - P_s)$,   [34]

where K_p is the proportionality constant and equals A_1d_1/A_2d_2.

Figure 10.31 shows the basic form of a proportional plus integral pneumatic controller. A difference in pressure between the process and set-point bellows results in a movement of the flapper. This changes the output pressure and the pressure in the proportional bellows. The result is a motion of the flapper until the turning moment resulting from the force exerted by the proportional bellows balances that resulting from the difference between the process and set-point bellows. While this is happening the integral bellows has barely been affected because of the time delay introduced by the restriction. However as the integral bellows slowly comes

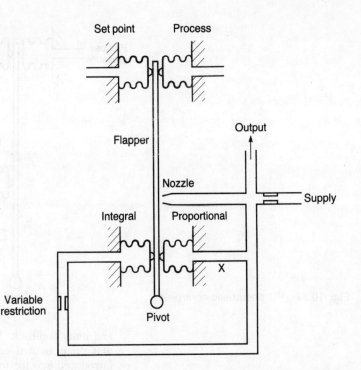

Fig. 10.31 PI pneumatic controller

up to the output pressure it moves the flapper and so changes the output pressure.

Figure 10.32 shows the basic form of a proportional plus derivative pneumatic controller. The restriction in the air supply to the proportional bellows means that it cannot respond quickly to air pressure changes. Thus when the flapper moves as a consequence of a pressure difference between the set-point bellows and the process bellows the air escaping from the nozzle changes. The consequential change in the output pressure is a rather rapid change since the restriction prevents the proportional bellows from responding quickly. The result is a change which is proportional to the rate of change of the pressure difference between the set-point and process bellows. With time the proportional bellows responds and a further change takes place which is proportional to the pressure difference between the set-point and process bellows. The restriction in the line to the proportional bellows is generally variable since it determines the value of K_D.

Figure 10.32 becomes a PID controller if another variable constriction is introduced at point X, such a constriction introducing the integral element referred to in Fig. 10.31.

Problems

1 A feedback system has a forward-path transfer function of

$$\frac{1}{s(s^2 + 3s + 5)}$$

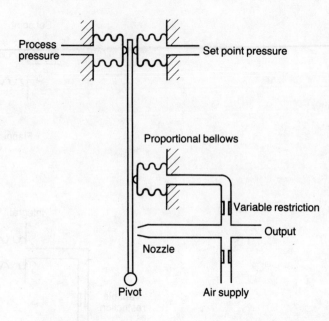

Fig. 10.32 PD pneumatic controller

and unity feedback. What will be the steady-state error with a unit ramp input if (a) a proportional controller with gain 4 is introduced into the forward path, (b) an integral controller with time constant 2s is used instead of the proportional controller?

2 What open-loop zeros and roots are introduced into a feedback system if a PI controller, with proportional gain 4 and integral time constant 2s, is introduced into the forward path?

3 A feedback system has an open-loop transfer function of

$$G_o(s) = \frac{s+3}{s(s+1)(s+2)}$$

By how much will the point of intersection of the root loci asymptotes with the real axis be moved if a PI controller is introduced with an integral time constant of 5s and a proportional gain of 2?

4 A feedback system has proportional plus derivative control with a plant having a forward-path transfer function of

$$\frac{1}{s(s+2)}$$

Select the values of K_p and K_d so that the system has a natural angular frequency of 0.5 rad/s and a damping factor of 0.7.

5 A PID control system is used to control a plant having a transfer function of

$$\frac{1}{s(s+3+j1)(s+3-j1)}$$

If the PID controller has an integral time constant of 4s and a derivative time constant of 1s, what will be the open-loop roots and zeros of the control system?

6 Determine the settings of K_p, K_i and K_d required for a three-

mode controller which gives a process reaction curve with a lag L of 200s and a gradient R of 0.010%/s when the test signal was a 5% change to the correction unit input.

7 When tuning a three-mode controller by the ultimate cycle method it was found that oscillations began when the proportional band was increased to 20% with the oscillations having a periodic time of 200s. What are suitable values for K_p, K_i and K_d?

8 Describe the effects on a control system of including (a) proportional control, (b) integral control, (c) proportional plus integral control, (d) proportional plus derivative control, (e) proportional plus integral plus derivative control.

9 Explain what is meant by velocity feedback and explain the reasons for using it.

10 What is the maximum value of the gain K which will give stable conditions for the system shown in Fig. 10.33 with velocity and position feedback?

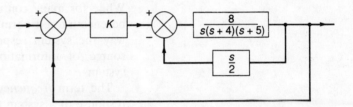

Fig. 10.33 Problem 10

11 What is the effect on the intercept of the asymptotes with the real axis of the root locus diagram, and hence the relative stability, of introducing (a) a lead compensator of transfer function $(s + 2)/(s + 3)$, (b) a lag compensator of transfer function $(s + 3)/(s + 2)$ into the forward path of a feedback system having an open-loop transfer function of $1/[s(s + 1)]$?

12 Sketch the root locus diagrams for a feedback system having an open-loop transfer function of

$$G_o(s) = \frac{K}{(s + 3)(s + 5)}$$

when (a) uncompensated, (b) with a lead compensator having a transfer function of $(s + 1)/(s + 2)$, (c) with a lag compensator having a transfer function of $(s + 2)/(s + 1)$.

13 Explain how operational amplifiers can be used to provide proportional, integral and derivative forms of control.

11 Frequency response

Introduction

In Chapter 6 the output of systems when subject to an impulse, step or ramp input was considered. This chapter extends this consideration to when there is a sinusoidal input. While for many control systems a sinusoidal input might not be encountered normally it is a useful testing input since the way the system responds to such an input is a very useful source of information to aid the design and analysis of systems.

The term *frequency response* is defined as the steady-state response of a system to a sinusoidal input, the response being monitored over a range of frequencies. The steady-state response is the response that remains after all transients have decayed to zero. There are several techniques that are used for the analysis of frequency response data. In this chapter the two techniques considered are those of Bode and Nyquist.

Frequency response

If a sinusoidal input is applied to a linear system the output is also sinusoidal and of the same frequency. The output can differ from the input in amplitude and in phase. The ratio of the amplitude of the output to the amplitude of the input is usually referred to as the *magnitude*, although sometimes it is called the amplitude ratio or gain. The shift of phase of the sinusoidal output relative to that of the sinusoidal input is called the *phase*. The variation of the magnitude and phase with frequency is called the *frequency response* of the system.

Transfer function

The transfer function $G(s)$ of a system can in general be represented by

$$G(s) = \frac{K(s - z_1)(s - z_2) \ldots (s - z_m)}{(s - p_1)(s - p_2) \ldots (s - p_n)} \quad [1]$$

where K is the gain, $z_1, z_2, \ldots z_m$ the zeros of the system and

p_1, p_2, . . . p_n the poles, there being m zeros and n poles. Thus, since $G(s)$ is the ratio of the output to input, i.e. $G(s) = \theta_o(s)/\theta_i(s)$, then the output is given by

$$\theta_o(s) = \frac{K(s - z_1)(s - z_2) \ldots (s - z_m)}{(s - p_1)(s - p_2) \ldots (s - p_n)} \theta_i(s) \qquad [2]$$

Thus if we consider a sinusoidal input

$$\theta_i = a \sin \omega t$$

where a is the amplitude of the input and ω is the angular frequency in rad/s, then

$$\theta_i(s) = \frac{a\omega}{s^2 + \omega^2}$$

and equation [2] becomes

$$\theta_o(s) = \frac{K(s - z_1)(s - z_2) \ldots (s - z_m)}{(s - p_1)(s - p_2) \ldots (s - p_n)} \frac{a\omega}{s^2 + \omega^2} \qquad [3]$$

We can follow through the solution of this, using partial fractions, and obtain a relationship of the form

$$\theta_o(s) = \text{transient terms} + \text{steady-state terms}$$

The transient terms die away with time. Thus if we are only concerned with the steady state, the solution obtained is

$$\theta_o = a|G(j\omega)| \sin(\omega t + \phi) \qquad [4]$$

The steady-state output is sinusoidal with the same angular frequency ω as the input. $|G(j\omega)|$ is the magnitude of the transfer function $G(s)$ when s is replaced by $j\omega$. The function $G(j\omega)$, which is obtained by replacing s by $j\omega$ in $G(s)$, is called the *frequency response function*. In the same way as we talk of $G(s)$ being in the s domain (see Chapter 4) so we can talk of $G(j\omega)$ being the transfer function $G(s)$ in the *frequency domain*. $G(j\omega)$ can be found by replacing every value of s in $G(s)$ by $j\omega$ and then rearranging the expression to obtain it in a form which enables the real and imaginary parts to be separated, hence identifying the magnitude and the phase.

Consider the transfer function

$$G(s) = \frac{1}{s + 2}$$

If we let $s = j\omega$ then

$$G(j\omega) = \frac{1}{j\omega + 2}$$

If we multiply the top and bottom of the expression by $(-j\omega + 2)$ then

$$G(j\omega) = \frac{-j\omega + 2}{\omega^2 + 4} = \frac{2}{\omega^2 + 4} - \frac{j\omega}{\omega^2 + 4}$$

This equation gives the frequency transfer function as a complex number in the form $x + jy$. Hence, the magnitude $|G(j\omega)|$ is (see note at the end of this section)

$$|G(j\omega)| = \sqrt{\left[\left(\frac{2}{\omega^2 + 4}\right)^2 + \left(\frac{\omega}{\omega^2 + 4}\right)^2\right]}$$

$$|G(j\omega)| = \frac{1}{\sqrt{(\omega^2 + 4)}}$$

and the phase ϕ is given by

$$\tan\phi = \left[\frac{\omega/(\omega^2 + 4)}{2/(\omega^2 + 4)}\right] = -\frac{\omega}{2}$$

Because the y term is negative and the x term positive, then $\tan\phi$ is negative and so ϕ is the angle by which the output lags behind the input.

The following is a brief reminder about complex numbers. A complex quantity can be represented by $(x + jy)$, where x is the real part and y the imaginary part of the complex number. On a graph with the imaginary component as the y-axis and the real part as the x-axis then the x and y are the Cartesian co-ordinates of the point representing the complex number (Fig. 11.1). Alternatively, it can be represented in polar form (see Chapter 9) as $r(\cos\phi + j\sin\phi)$, where on the graph of imaginary component against real component r is the length of of the line joining the origin to the point representing the complex number and ϕ is the angle between the line and the x-axis. The term $(\cos\phi + j\sin\phi)$ can be represented by $\angle\phi$ and so the complex number by $r\angle\phi$, where r is the magnitude and ϕ the phase of the complex number. Thus using Pythagoras' theorem the magnitude is given by

$$r = \sqrt{(x^2 + y^2)} \tag{5}$$

and phase ϕ by

$$\tan\phi = \frac{y}{x} \tag{6}$$

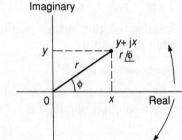

Imaginary

Fig. 11.1 A complex number

The signs of the y and x terms must be taken into account in determining $\tan\phi$. Positive y and positive x would mean that ϕ was between 0 and 90°, positive y and negative x that ϕ was between 90° and 180°, negative y and negative x that ϕ was between 180° and 270° and negative y and positive x that ϕ was between 270° and 360°.

Example 1

What are the steady-state magnitude and the phase of the output from a system when subject to a sinusoidal input of $\theta_i = 2\sin(3t + 60°)$ if it has a transfer function of

$$G(s) = \frac{4}{s + 1}$$

Answer

Replacing s by $j\omega$ gives

$$G(j\omega) = \frac{4}{j\omega + 1}$$

Multiplying top and bottom of the equation by $(-j\omega + 1)$ gives

$$G(j\omega) = \frac{-j4\omega + 4}{\omega^2 + 1} = \frac{4}{\omega^2 + 1} - \frac{j4\omega}{\omega^2 + 1}$$

Hence

$$|G(j\omega)| = \frac{4}{\sqrt{(\omega^2 + 1)}}$$

and the phase angle is given by

$$\tan\phi = \frac{y}{x} = -\omega$$

and, because y is negative and x positive, is the angle by which the output lags behind the input. For the specified input $\omega = 3$ rad/s. Thus

$$|G(j\omega)| = \frac{4}{\sqrt{(3^2 + 1)}} = 1.3$$

and

$$\tan\phi = -\omega = -3$$

Thus $\phi = -72°$. The output is thus

$$\theta_o = a|G(j\omega)|\sin(\omega t + \theta + \phi)$$

where a is the input amplitude 2, ω is the input angular frequency 3 rad/s, θ is the phase angle of the input and ϕ is the difference in phase angles between the output and the input. Hence

$$\theta_o = 2.6\sin(3t - 12°)$$

Example 2

For a system having the transfer function

$$G(s) = \frac{3}{s + 2}$$

determine (*a*) the magnitude and phase of the frequency response and (*b*) produce a table showing the magnitude and phase values with angular frequency for $\omega = 0, 2, 10, 100, \infty$ rad/s.

Answer

(*a*) Replacing *s* by jω gives

$$G(j\omega) = \frac{3}{j\omega + 2}$$

Multiplying top and bottom of the equation by $(-j\omega + 2)$ gives

$$G(j\omega) = \frac{-j3\omega + 6}{\omega^2 + 4} = \frac{6}{\omega^2 + 4} - \frac{j3\omega}{\omega^2 + 4}$$

The equation is now in the form $x + jy$. Hence, using equation [5]

$$\text{Magnitude} = r = \sqrt{(x^2 + y^2)}$$

$$|G(j\omega)| = \frac{3}{\sqrt{(\omega^2 + 4)}}$$

The phase angle is given by equation [6] as

$$\tan\phi = \frac{y}{x} = -\frac{\omega}{2}$$

and, because *y* is negative and *x* positive, this is the angle by which the output lags behind the input.

(*b*) When ω = 0 then $|G(j\omega)| = 3/\sqrt{4} = 1.5$. When ω = 2 then $|G(j\omega)| = 3/\sqrt{8} = 1.06$. When ω = 10 then $|G(j\omega)| = 3/\sqrt{104} = 0.29$. When ω = 100 then $|G(j\omega)| = 3/\sqrt{10004} = 0.03$. When ω = ∞ then $|G(j\omega)| = 3/\infty = 0$.

When ω = 0 then $\tan\phi = 0$ and so $\phi = 0°$. When ω = 2 then $\tan\phi = -1$ and so $\phi = -45°$. When ω = 10 then $\tan\phi = -5$ and $\phi = 78.7°$. When ω = 100 then $\tan\phi = -50$ and $\phi = -88.9°$. When ω = ∞ then $\tan\phi = -\infty$ and $\phi = -90°$.

ω	0	2	10	100	∞
$\|G(j\omega)\|$	1.5	1.06	0.29	0.03	0
φ	0	−45°	−78.7°	−88.9°	−90°

Frequency response for a first-order system   A first-order system has a transfer function of the form

$$G(s) = \frac{1}{1 + \tau s} \tag{7}$$

where τ is the time constant. The frequency-response function $G(j\omega)$ can be obtained by replacing *s* by jω. Hence

$$G(j\omega) = \frac{1}{1 + j\omega\tau} \tag{8}$$

Multiplying the top and bottom of the expression by $(1 - j\omega\tau)$ gives

$$G(j\omega) = \frac{1 - j\omega\tau}{1 + \omega^2\tau^2} = \frac{1}{1 + \omega^2\tau^2} - \frac{j\omega\tau}{1 + \omega^2\tau^2}$$

This is of the form $(x + jy)$ and so the magnitude $|G(j\omega)|$ is, by equation [5],

$$\text{Magnitude} = r = \sqrt{(x^2 + y^2)}$$

$$|G(j\omega)| = \frac{1}{\sqrt{(1 + \omega^2\tau^2)}} \qquad [9]$$

The phase ϕ is given by equation [6] as

$$\tan\phi = \frac{y}{x} = -\omega\tau \qquad [10]$$

The phase angle is the amount by which the output lags behind the input since the y term is negative and the x term positive.

Example 3

The transfer function for a system (an electrical circuit with a resistor in series with a capacitor across which the output is taken) is

$$G(s) = \frac{1}{RCs + 1}$$

What is (a) the frequency response function $G(j\omega)$, and (b) the magnitude and phase of that response?

Answer

(a) The system is first order with a time constant τ of RC. Thus the frequency-response function will be of the form given by equation [8], i.e. the transfer function with s replaced by $j\omega$.

$$G(j\omega) = \frac{1}{1 + j\omega RC}$$

(b) The magnitude is thus of the form given by equation [9]

$$|G(j\omega)| = \frac{1}{\sqrt{(1 + \omega^2 R^2 C^2)}}$$

The phase ϕ is given by equation [10] as

$$\tan\phi = -\omega RC$$

Frequency response for a second-order system

A second-order system has a transfer function of the form

$$G(s) = \frac{\omega_n^2}{s^2 + 2\zeta\omega_n s + \omega_n^2} \qquad [11]$$

where ω_n is the natural angular frequency and ζ the damping ratio. The frequency response is obtained by replacing s by $j\omega$. Thus the frequency response function $G(j\omega)$ is given by

$$G(j\omega) = \frac{\omega_n^2}{-\omega^2 + j2\,\zeta\omega\omega_n + \omega_n^2} = \frac{\omega_n^2}{(\omega_n^2 - \omega^2) + j2\,\zeta\omega\omega_n}$$

$$= \frac{1}{[1 - (\omega/\omega_n)^2] + j2\,\zeta(w/\omega_n)}$$

Multiplying the top and bottom of the expression by

$$[1 - (\omega/\omega_n)^2] - j2\,\zeta(\omega/\omega_n)$$

gives

$$G(j\omega) = \frac{1 - (\omega/\omega_n)^2 - j2\,\zeta(\omega/\omega_n)}{[1 - (\omega/\omega_n)^2]^2 + [2\,\zeta(\omega/\omega_n)]^2}$$

This is of the form $x + jy$ and so the magnitude $|G(j\omega)|$ is given by equation [5] as

$$\text{Magnitude} = r = \sqrt{(x^2 + y^2)}$$

$$|G(j\omega)| = \frac{1}{\sqrt{\{[1 - (\omega/\omega_n)^2]^2 + [2\,\zeta(\omega/\omega_n)]^2\}}} \qquad [12]$$

The phase ϕ is given by equation [6] as

$$\tan\phi = \frac{y}{x} = -\frac{2\,\zeta(\omega/\omega_n)}{1 - (\omega/\omega_n)^2} \qquad [13]$$

The minus sign indicates that the output lags behind the input.

Frequency response from a pole-zero plot

The magnitude and phase of $G(j\omega)$ can be found from the pole-zero plot for a system. Suppose we have a system with a transfer function

$$G(s) = \frac{1}{s + 1}$$

Such a system has a pole at $s = -1$ (Fig. 11.2). If the input to such a system is sinusoidal then $s = j\omega$. This defines a point on the $j\omega$ axis according to the value of the input angular frequency ω. In Fig. 11.2 ω has been chosen to have the value 1 rad/s. The transfer function thus becomes

$$G(j\omega) = \frac{1}{j1 + 1} = \frac{1}{\sqrt{2}\angle 45°}$$

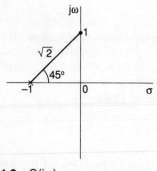

Fig. 11.2 $G(j\omega)$

But $\sqrt{2}$ is the length of the line drawn joining the pole to the $s = j\omega$ point and $\angle 45°$ is the angle the line makes with the axis. Thus $G(j\omega)$ is the reciprocal of this, i.e. $(1/\sqrt{2})\angle -45°$.

In general where the transfer function has a number of zeros and poles, i.e.

$$G(s) = \frac{K(s + z_1)(s + z_2) \dots (s + z_m)}{(s + p_1)(s + p_2) \dots (s + p_n)}$$

then the procedure is:

1 Plot the positions of each pole and zero.
2 Mark the position $s = j\omega$.
3 Draw lines from each pole and each zero to the point $s = j\omega$.
4 Measure the lengths and angles of each of the lines.
5 The frequency-response function is then

$$|G(j\omega)|$$

$$= \frac{K \times \text{product of the lengths of the lines from zeros}}{\text{product of the lengths of the lines from poles}} \quad [14]$$

$$\angle G(j\omega)$$

$$= \text{Sum of angles of lines from zeros}$$

$$- \text{sum of angles of lines from poles} \quad [15]$$

Example 4

A system has a transfer function

$$G(s) = \frac{s(s + 2)}{s^2 + 2s + 2}$$

Determine the output from the system when there is an input to it of $10 \sin 2t$.

Answer

The transfer function can be written as

$$G(s) = \frac{(s + 0)(s + 2)}{(s + 1 + j1)(s + 1 - j1)}$$

Figure 11.3 shows the pole-zero plot, there being zeros at $s = 0$ and $s = -2$ and poles at $s = 1 + j1$ and $s = 1 - j1$.

The input has an angular frequency of 2 and thus $j\omega$ is $+2$. Lines are drawn from this point to each of the poles and zeros. The zeros give lines with magnitude 2, angle 90° and magnitude $\sqrt{8}$, angle 45° and the poles give magnitude $\sqrt{2}$, angle 45° and magnitude $\sqrt{10}$, angle 71.6°. Thus

$$G(j\omega) = \frac{(2 \times \sqrt{8})}{(\sqrt{2} \times \sqrt{10})} \angle (90° + 45° - 45° - 71.6°)$$

$$= 1.26 \angle 18.4°$$

Thus, since $G(j\omega) = $ output/input for a sinusoidal signal, then since the input is $10 \angle 0°$

$$\text{Output} = 1.26 \angle 18.4° \times 10 \angle 0° = 12.6 \angle 18.4°$$

$$= 12.6 \sin (2t + 18.4°)$$

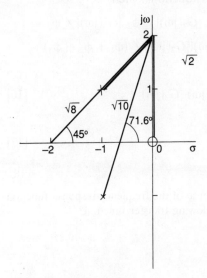

Fig. 11.3 Example 4

Frequency response for series elements

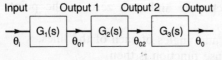

Fig. 11.4 Elements in series

If a system consists of a number of elements in series, as in Fig. 11.4, then the transfer function $G(s)$ of the system is the product of the transfer functions of the series elements, i.e.

$$G(s) = G_1(s)G_2(s)G_3(s) \ldots \text{etc.}$$

Hence, for the frequency-response function when s is replaced by $j\omega$,

$$G(j\omega) = G_1(j\omega)G_2(j\omega)G_3(j\omega) \ldots \text{etc.}$$

Since $G_1(j\omega)$ can be represented by equation [10] as

$$G_1(j\omega) = |G_1(j\omega)| \angle \phi$$

and similarly for the others functions, then

$$G(j\omega) = |G_1(j\omega)| \angle \phi_1 \; |G_2(j\omega)| \angle \phi_2 \; |G_3(j\omega)| \angle \phi_3$$
$$= |G_1(j\omega)||G_2(j\omega)||G_3(j\omega)| \angle (\phi_1 + \phi_2 + \phi_3)$$

Thus

$$|G(j\omega)| = |G_1(j\omega)||G_2(j\omega)||G_3(j\omega)| \ldots \qquad [16]$$

and for the phases

$$\phi = \phi_1 + \phi_2 + \phi_3 \ldots \qquad [17]$$

Example 5

What is the magnitude and phase of the frequency-response function of a system which has the following transfer function?

$$G(s) = \frac{1}{(s + 1)(2s + 1)}$$

Answer

This system can be considered to consist of two elements in series, one with a transfer function of $1/(s + 1)$ and the other with $1/(2s + 1)$. Each system is a first-order system and thus the general results derived earlier for such a form of system can be used for each element independently, or the response of each system derived independently from first principles. For

$$G_1(s) = \frac{1}{s + 1}$$

$$G_1(j\omega) = \frac{1}{j\omega + 1} = \frac{-j\omega + 1}{\omega^2 + 1}$$

Hence

$$|G_1(j\omega)| = \frac{1}{\sqrt{(\omega^2 + 1)}}$$

and

$$\tan \phi_1 = -\omega$$

$$\phi = -\tan^{-1} \omega$$

For the second system

$$G_2(s) = \frac{1}{2s + 1}$$

Hence

$$G_2(j\omega) = \frac{1}{2j\omega + 1} = \frac{-2j\omega + 1}{4\omega^2 + 1}$$

and so

$$|G_2(j\omega)| = \frac{1}{\sqrt{(4\omega^2 + 1)}}$$

and

$$\tan \phi = -2\omega$$
$$\phi = -\tan^{-1} 2\omega$$

Thus for the two elements in series, equation [16] gives

$$|G(j\omega)| = \left[\frac{1}{\sqrt{(\omega^2 + 1)}}\right]\left[\frac{1}{\sqrt{(4\omega^2 + 1)}}\right]$$

and equation [17] gives

$$\phi = -\tan^{-1}\omega - \tan^{-1} 2\omega$$

Bode plots

The *Bode plot* consists of two graphs, one of the magnitude plotted against the frequency and one of the phase angle plotted against frequency. The magnitude and the frequency are plotted using logarithmic scales.

For a system having a transfer function involving a number of terms, equation [16] indicates that the resultant magnitude is the product of the magnitudes of the constituent elements, i.e.

$$|G(j\omega)| = |G_1(j\omega)||G_2(j\omega)||G_3(j\omega)| \ldots$$

Taking logarithms to base 10, this equation becomes

$$\lg|G(j\omega)| = \lg|G_1(j\omega)| + \lg|G_2(j\omega)| + \lg|G_3(j\omega)| \qquad [18]$$

Thus plotting a graph of $\lg|G(j\omega)|$ against frequency means that we can just add the contributions due to the individual magnitude terms. For example, if we wanted to obtain the Bode plot for

$$G(j\omega) = \frac{5(1 + j\omega)}{2 + j\omega}$$

then we can plot separately the log graphs for magnitudes of element 5, of element $(1 + j\omega)$ and of element $1/(2 + j\omega)$ and just add them to obtain the plot for $|G(j\omega)|$.

It is usual to express the magnitude in units of decibels (dB).

$$\text{Magnitude in dB} = 20\lg|G(\text{j}\omega)| \qquad [19]$$

Thus if $|G(\text{j}\omega)| = 2$ then, since $20\lg 2$ is 20 the magnitude is 20 dB.

The phase graph when there are a number of elements is just the sum of the separate elements (equation [19]). The frequency scale used for both the magnitude and the phase plots is logarithmic. This is to enable a graph to cover a greater range of frequencies and also because it often leads to straight-line graphs.

Because Bode plots for a system can be built up from the plots for the individual elements within the transfer function for that system it is useful to consider the plots for elements commonly found in transfer functions. By using such elements the Bode plot for a wide range of systems can rapidly be built up. The basic elements considered are as follows.

Constant gain

This is where

$$G(s) = K \qquad [20]$$

and thus

$$G(\text{j}\omega) = K \qquad [21]$$

For such a system the magnitude is, in decibels,

$$|G(\text{j}\omega)| = 20\lg K \qquad [22]$$

and the phase is zero. The Bode plot is thus of the form shown in Fig. 11.5. The magnitude plot is a line of constant magnitude. Changing the gain K merely shifts the magnitude plot up or down by a certain number of decibels.

A pole at the origin

This is where

$$G(s) = \frac{1}{s} \qquad [23]$$

and so

$$G(\text{j}\omega) = \frac{1}{\text{j}\omega} = -\frac{\text{j}}{\omega} \qquad [24]$$

For such a system the magnitude is, in decibels,

$$|G(\text{j}\omega)| = 20\lg(1/\omega) = -20\lg\omega \qquad [25]$$

When $\omega = 1$ rad/s then $|G(\text{j}\omega)| = 0$, and when $\omega = 10$ rad/s then $|G(\text{j}\omega)| = -20$ dB. For each tenfold increase in frequency

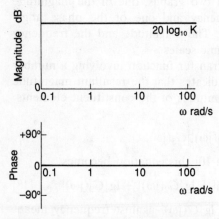

20 log₁₀ K

Fig. 11.5 Bode plot for constant gain

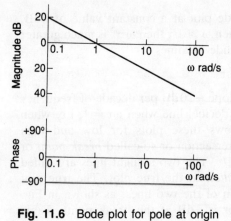

Fig. 11.6 Bode plot for pole at origin

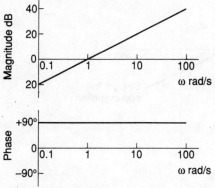

Fig. 11.7 Bode plot for zero at origin

the magnitude drops by $-20\,\text{dB}$. The Bode magnitude plot is thus a straight line of slope $-20\,\text{dB}$ per decade of frequency which passes through $0\,\text{dB}$ at $\omega = 1\,\text{rad/s}$ (Fig. 11.6). The phase of such a system is given by

$$\tan\phi = \frac{(-1/\omega)}{0} = -\infty$$

Hence ϕ is constant, for all frequencies, at $-90°$.

A zero at origin

This is where

$$G(s) = s \tag{26}$$

and thus

$$G(j\omega) = j\omega \tag{27}$$

The magnitude, in decibels, is thus $20\lg\omega$. Thus when $\omega = 1\,\text{rad/s}$ then $|G(j\omega)| = 0\,\text{dB}$, and when $\omega = 10\,\text{rad/s}$ then $|G(j\omega)| = 20\,\text{dB}$. The magnitude is a Bode plot with a straight line of slope $+20\,\text{dB}$ per decade of frequency and which passes through $0\,\text{dB}$ at $\omega = 1\,\text{rad/s}$ (Fig. 11.7). The phase is given by

$$\tan\phi = \frac{\omega}{0} = +\infty$$

Hence the phase is constant at $90°$ regardless of frequency.

A real pole

This means a first-order lag system, where

$$G(s) = \frac{1}{\tau s + 1} \tag{28}$$

and thus

$$G(j\omega) = \frac{1}{j\omega\tau + 1} = \frac{1 - j\omega\tau}{1 + \omega^2\tau^2} \tag{29}$$

The magnitude, in decibels, is

$$20\lg\left[\frac{1}{\sqrt{(1 + \omega^2\tau^2)}}\right]$$

and the phase angle

$$\tan\phi = -\omega\tau$$

When $\omega \ll 1/\tau$ then $\omega^2\tau^2$ is negligible compared with 1 and so the magnitude is thus $0\,\text{dB}$. Hence at low frequencies there is a straight-line magnitude plot at a constant value of 0 dB. For higher frequencies when $\omega \gg 1/\tau$ then $\omega^2\tau^2$ is much greater than 1 and so the magnitude becomes

$$20 \lg (1/\omega\tau) = -20 \lg \omega\tau$$

This is a straight-line of slope -20 dB per decade of frequency which intersects the zero decibel line when $\omega\tau = 1$, i.e. when $\omega = 1/\tau$. Figure 11.8 shows these plots for low and high frequencies, with their intersection or so-called *break point* or *corner frequency* at $\omega = 1/\tau$. The two straight lines are called the *asymptotic approximation* to the true plot. The true plot rounds off the intersection of the two lines, as shown in Fig. 11.8. The maximum error is 3 dB at the break point. Table 11.1 gives the differences between the true values and the asymptotic values.

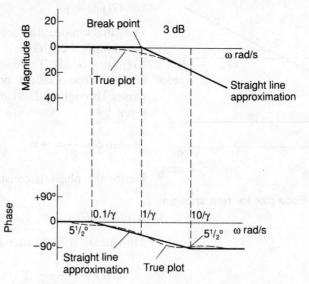

Fig. 11.8 Bode plot for a real pole

The phase angle is $-\tan^{-1}\omega\tau$. At low frequencies when ω is less than about $0.1/\tau$ then the phase is virtually $0°$. At high frequencies when ω is more than $10/\tau$ then the phase is virtually $-90°$. Between these two extremes the phase angle can be considered to give a reasonable straight line, as in Fig. 11.8. This line is the asymptotic approximation. The maximum

Table 11.1 Asymptotic errors for a real pole, or real zero

ω	$0.10/\tau$	$0.2/\tau$	$0.50/\tau$	$1/\tau$	$2/\tau$	$5/\tau$	$10/\tau$
Magnitude error (dB)	-0.04	-0.02	-1.0	-3.0	-1.0	-0.2	-0.04
Phase error	$-5.7°$	$+2.3°$	$+4.9°$	$0°$	$-4.9°$	$-2.3°$	$+5.7°$

Note: true value = linear approximation + error

error is assuming a straight line is $5\frac{1}{2}°$. At $\omega = 1/\tau$, the break point, the phase is $-45°$. Table 11.1 gives the differences between the true values and the asymptotic values.

A real zero

This means a first-order lead system where

$$G(s) = 1 + \tau s \tag{30}$$

and thus

$$G(j\omega) = 1 + j\omega\tau \tag{31}$$

The magnitude, in decibels, is thus

$$20 \lg \sqrt{(1 - \omega^2\tau^2)}$$

and the phase

$$\tan\phi = \omega\tau$$

At low frequencies when $\omega \ll 1/\tau$ then the $\omega^2\tau^2$ term is insignificant in comparison with 1 and so the magnitude is a straight line of 0 dB. At high frequencies when $\omega \gg 1/\tau$ then 1 is insignificant in comparison with the $\omega^2\tau^2$ term and so the magnitude is $20 \lg \omega\tau$ and hence a straight line of slope 20 dB per decade of frequency with a break point at $\omega = 1/\tau$. Figure 11.9 shows how these two straight lines approximate to the true magnitude plot, the maximum error being 3 dB at the break point. See Table 11.1 for the errors between true values and the asymptotic lines.

The phase angle is $\tan^{-1}\omega\tau$. At low frequencies when ω is less than about $0.1/\tau$ then the phase is virtually 0°. At high frequencies when ω is more than $10/\tau$ then the phase is virtually 90°. Between these two extremes the phase angle can be considered to give a reasonable straight line, as in Fig. 11.9. The maximum error is assuming a straight line is $5\frac{1}{2}°$, see Table

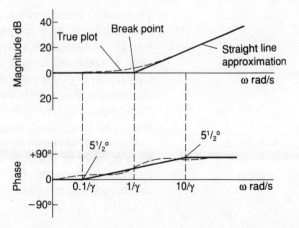

Fig. 11.9 Bode plot for a real zero

11.1 for more details. At $\omega = 1/\tau$, the break point, the phase is 45°.

A pair of complex poles

This is where

$$G(s) = \frac{\omega_n^2}{s^2 + 2\zeta\omega_n s + \omega_n^2} \qquad [32]$$

and thus

$$G(j\omega) = \frac{\omega_n^2}{-\omega^2 + j2\zeta\omega_n + \omega_n^2}$$

$$G(j\omega) = \frac{1}{[1 - (\omega/\omega_n)^2] + j[2\zeta(\omega/\omega_n)]} \qquad [33]$$

Hence we can write

$$G(j\omega) = \frac{[1 - (\omega/\omega_n)^2] - j[2\zeta(\omega/\omega_n)]}{[1 - (\omega/\omega_n)^2]^2 + [2\zeta(\omega/\omega_n)]^2}$$

and so the magnitude, in decibels, is

$$|G(j\omega)| = 20\lg \sqrt{\left\{ \frac{1}{[1 - (\omega/\omega_n)^2]^2 + [2\zeta(\omega/\omega_n)]^2} \right\}}$$

$$|G(j\omega)| = -20\lg \sqrt{\{[1 - (\omega/\omega_n)^2]^2 + [2\zeta(\omega/\omega_n)]^2\}} \qquad [34]$$

and the phase is

$$\tan\phi = -\frac{2\zeta(\omega/\omega_n)}{1 - (\omega/\omega_n)^2} \qquad [35]$$

For $(\omega/\omega_n) \ll 1$ then the magnitude approximates to

$$|G(j\omega)| \approx -20\log_{10} 1 = 0\,dB$$

For $(\omega/\omega_n) \gg 1$ then the magnitude approximates to

$$|G(j\omega)| \approx -20\lg(\omega/\omega_n)^2$$

Thus at low frequencies the magnitude plot is a straight line at 0 dB, while at high frequencies it is a straight line of slope −40 dB per decade of frequency. The intersection or break point of these two lines is at $\omega = \omega_n$, this being the value of the high-frequency expression which gives 0 dB and so the intersection point with the 0 dB line. The magnitude plot is thus approximately given by these two straight asymptotic lines. The true value, however, depends on the damping ratio ζ. Figure 11.10 shows the two straight lines and the true plots for a number of different damping ratios. Table 11.2 gives the differences between the true and asymptotic values for a number of damping ratios.

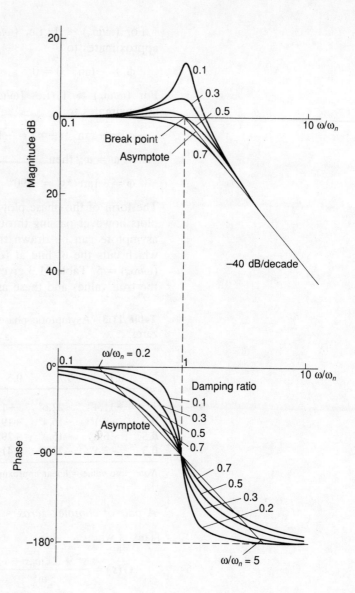

Fig. 11.10 Bode plot for a pair of complex poles

Table 11.2 Asymptotic magnitude errors in dB for pair of complex poles or zeros

ζ	0.1	0.2	0.5	1.0	2	5	10
				ω/ω_n			
1	−0.09	−0.34	−1.92	−6.0	−1.92	−0.34	−0.09
0.7	0	−0.01	−0.26	−3.0	−0.26	−0.01	0
0.5	+0.04	+0.17	+0.90	0	+0.90	+0.17	+0.04
0.3	+0.07	+0.29	+1.85	+4.4	+1.85	+0.29	+0.07

Note: true value = linear approximation + error

For $(\omega/\omega_n) \ll 1$, i.e. $(\omega/\omega_n) = 0.2$ or less, then the phase approximates to

$$\phi \approx -\tan^{-1}0 = 0°$$

For $(\omega/\omega_n) \gg 1$, i.e. $(\omega/\omega_n) = 5$ or more, then the phase approximates to

$$\phi \approx -\tan^{-1}(-\infty) = -180°$$

When $\omega = \omega_n$ then

$$\phi = -\tan^{-1}\infty = -90°$$

The form of the phase plot depends on the damping ratio, all plots however passing through the $-90°$ angle at $\omega = \omega_n$. An asymptote can be drawn through the $-90°$ angle at $\omega = \omega_n$ which cuts the $0°$ line at $(\omega/\omega_n) = 0.2$ and the $-180°$ line at $(\omega/\omega_n) = 5$. Table 11.3 gives the error, or difference, between the true values and these asymptotic values.

Table 11.3 Asymptotic phase errors for pair of complex poles or zeros

ζ	0.1	0.2	0.5	1.0	2	5	10
				$\omega/\omega n$			
1	$-11.4°$	$-22.6°$	$+1.6°$	0	$-1.6°$	$+22.6°$	$+11.4°$
0.7	$-8.1°$	$-16.4°$	$+19.6°$	0	$-19.6°$	$+16.4°$	$+8.1°$
0.5	$-5.8°$	$-15.3°$	$+29.2°$	0	$-29.2°$	$+15.3°$	$+5.8°$
0.3	$-3.5°$	$-22.3°$	$+41.1°$	0	$-41.1°$	$+22.3°$	$+3.5°$

Note: true value = linear approximation + error

A pair of complex zeros

This is where

$$G(s) = \frac{s^2 + 2\zeta\omega_n s + \omega_n^2}{\omega_n^2} \qquad [36]$$

Thus

$$G(j\omega) = \frac{-\omega^2 + j2\zeta\omega_n\omega + \omega_n^2}{\omega_n^2}$$

$$G(j\omega) = [1 - (\omega/\omega_n)^2] + j[2\zeta(\omega/\omega_n)] \qquad [37]$$

Hence the magnitude, in decibels, is

$$|G(j\omega)| = \sqrt{\{[1 - (\omega/\omega_n)^{-2}]^2 + [2\zeta(\omega/\omega_n)]^2\}} \qquad [38]$$

and the phase is

$$\tan\phi = \frac{2\zeta(\omega/\omega_n)}{1 - (\omega/\omega_n)^2} \qquad [39]$$

The magnitude differs only from that for the complex poles (equation [34]) in being positive while the latter is negative. Thus the magnitude plot is just the mirror image, about the 0 dB line, of the plot given in Fig. 11.10. The phase differs only from that for the complex poles (equation [35]) in being positive while the latter is negative. Thus the phase plot is just the mirror image, about the 0° line, of the plot given in Fig. 11.10. See Tables 11.2 and 11.3 for the differences between the true values and the asymptotes.

Example 6

Draw the asymptotes of the Bode diagram for a system having a transfer function of

$$G(s) = \frac{10}{2s + 1}$$

Answer

The transfer function is made up of two components: an element with transfer function 10 and one with transfer function $1/(2s + 1)$. The Bode plots can be drawn for each of these and then added together to give the required plot.

A system with a transfer function of $G(s) = 10$ is a constant-gain system. The plot will thus be like that in Fig. 11.5. The constant magnitude, in decibels, is $20 \lg 10 = 20$. The constant phase is 0°. The plot is shown in Fig. 11.11.

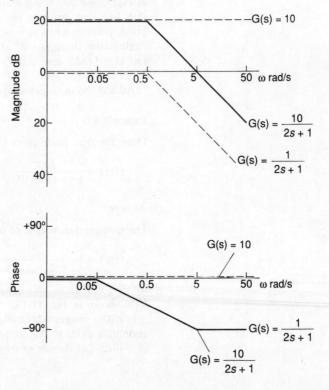

Fig. 11.11 Example 6

A system with a transfer function $G(s) = 1/(2s + 1)$ is one with a real pole and no zeros. The system is thus like the one that gave Fig. 11.8, the time constant τ being 2 s. The plot will thus have a break point at $1/\tau = 1/2 = 0.5$ s and the asymptotes will be as shown in Fig. 11.11.

Adding the two plots together gives the result shown in Fig. 11.11.

Example 7

Draw the asymptotes of the Bode diagram for a system having a transfer function of

$$G(s) = \frac{2.5}{s(s^2 + 3s + 25)}$$

Answer

The transfer function is made up of three components, one with a transfer function 0.1, one with $1/s$ and one with $25/(s^2 + 3s + 25)$.

The transfer function $G(s) = 0.1$ is one of constant gain and is thus like that shown in Fig. 11.5. The magnitude is a constant at $20 \lg 0.1 = -20 \lg_{10} 10 = -20$ dB. The phase is constant at $0°$. Figure 11.12 shows the plot.

The transfer function $G(s) = 1/s$ describes a system having a pole at the origin and is thus as shown in Fig. 11.6. The magnitude has a slope of -20 dB per decade of frequency and has the value 0 dB when $\omega = 1$ rad/s. The phase is constant at $-90°$. Figure 11.12 shows the plot.

The transfer function $G(s) = 25/(s^2 + 3s + 25)$ can be represented as $\omega_n^2/(s^2 + 2\zeta\omega_n + \omega_n^2)$ with $\omega_n = 5$ rad/s and $\zeta = 0.3$. The system is thus like that given in Fig. 11.10 for a pair of complex poles. The break point is when $\omega = \omega_n = 5$ rad/s. The asymptote for the phase angle passes through $-90°$ at the break point, being $0°$ at $(\omega/\omega_n) = 0.2$, i.e. $\omega = 1$ rad/s and $-180°$ at $(\omega/\omega_n) = 5$, i.e. $\omega = 25$ rad/s. Figure 11.12 shows the plot.

Adding the three plots together gives the required plot (Fig. 11.12).

Example 8

Draw the true Bode plots for a system with a transfer function of

$$G(s) = \frac{100}{s^2 + 6s + 100}$$

Answer

The transfer function is of the form

$$G(s) = \frac{\omega_n^2}{s^2 + 2\zeta\omega_n s + \omega_n^2}$$

with $\omega_n = 10$ rad/s and $\zeta = 0.3$. The Bode plots will thus be of the form shown in Fig. 11.10.

For the magnitude plot, the break point is 10 rad/s and the asymptote prior to this frequency has a slope of 0 and after it a slope of -40 dB per decade of frequency. The differences between the true

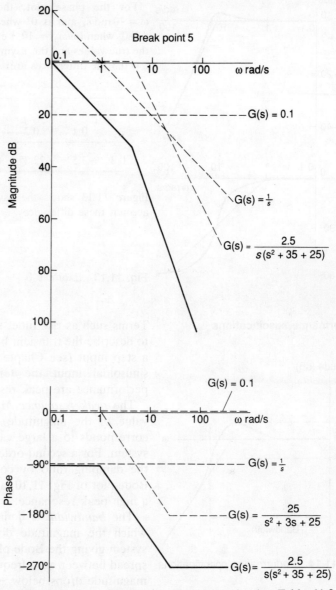

For this plot we see the asymptotes takes through 0 dB at a slope of 20 when $\log \omega = 1$, i.e. $\omega = 1$ rad, and is asymptotic to 20 + $\omega = 10$ rad. The difference between the true curve and the asymptote can be found using Table 11.2, Fig.

Fig. 11.12 Example 7

values and these asymptotes can be found using Table 11.2. For ζ = 0.3 the differences, in decibels, are:

ζ	ω/ωₙ						
	0.1	0.2	0.5	1.0	2	5	10
0.3	+0.07	+0.29	+1.85	+4.4	+1.85	+0.29	+0.07

Figure 11.13 shows the asymptotes and the true plot taking into account these differences.

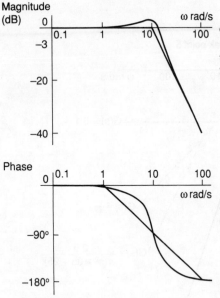

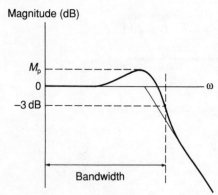

Performance specifications

Fig. 11.14 Performance specifications

For the phase plot, the asymptote passes through $-90°$ at $\omega = 10\,\text{rad/s}$, and is $0°$ when $(\omega/\omega_n) = 0.1$, i.e. $\omega = 1\,\text{rad/s}$ and is $-180°$ when $(\omega/\omega_n) = 10$, i.e. $\omega = 100\,\text{rad/s}$. The differences between the true values and this asymptote can be found using Table 11.3. For $\zeta = 0.3$ the differences are:

ζ	\multicolumn{7}{c}{ω/ω_n}
	0.1	0.2	0.5	1.0	2	5	10
0.3	$-3.5°$	$-22.3°$	$+41.1°$	0	$-41.1°$	$-22.3°$	$+3.5°$

Figure 11.13 shows the asymptote and the true plot taking into account these differences.

Fig. 11.13 Example 8

Terms such as rise time, settling time and overshoot are used to describe the transient behaviour of a system when subject to a step input (see Chapter 3). When a system is subject to a sinusoidal input the terms used to describe the system performance are peak resonance and bandwidth.

The *peak resonance* M_p is defined as being the maximum value of the magnitude (Fig. 11.14). A large value of M_p corresponds to a large value of the maximum overshoot of a system. For a second-order system it can be related directly to the damping ratio by comparison of the response with the Bode plot of Fig. 11.10, a low damping ratio corresponding to a high peak resonance.

The *bandwidth* is defined as the frequency band between which the magnitude does not fall below $-3\,\text{dB}$. For the system giving the Bode plot in Fig. 11.10 the bandwidth is the spread between zero frequency and the frequency at which the magnitude drops below $-3\,\text{dB}$. Since

$$20\lg G(j\omega) = -3 \qquad [40]$$

then $G(j\omega) = 0.707$, this being the cut off value for the frequency-response function.

Example 9

Estimate the bandwidth for the system giving the Bode plot in Fig. 11.13.

Answer

The $-3\,\text{dB}$ magnitude line crosses the true plot at about a quarter of a decade past 10 rad/s. This is about 11.7 rad/s. The bandwidth is thus from 0 rad/s up to 11.7 rad/s.

Using experimental frequency-response data

Frequency-response data can be determined experimentally for a system by using a sinusoidal signal as the input and monitoring the steady-state output in order to determine the ratio of the output amplitude to input amplitude (this then being the magnitude) and the phase difference between the output and the input. The measurement is repeated at a number of frequencies and hence the Bode plot obtained. Examination of the Bode plot can then lead to identification of the transfer function of the system.

The transfer function is usually determined from the magnitude plot with the phase plot being used as a check of the results. Asymptotes are drawn to the experimental data. If the slope of the initial asymptote is −20 dB/decade then there is a pole at the origin, if it is +20 dB/decade a zero at the origin. If the slope, in going from one asymptote to the next, changes by −20 dB/decade then there is a real pole and the transfer function includes $1/(\tau s + 1)$ where the frequency at the slope change point is $1/\tau$. If the slope changes by +20 dB/decade then there is a real zero and the transfer function includes $(\tau s + 1)$, where the frequency at the slope change point is $1/\tau$. If the slope, in going from one asymptote to the next, changes by −40 dB/decade there is a pair of complex poles and the transfer function has a term of the form

$$\frac{\omega_n^2}{s^2 + 2\zeta\omega_n s + \omega_n^2}$$

where ω_n is the frequency at the slope change. Determination of the difference between the magnitude plot and the asymptote break point enables the damping factor ζ to be estimated by means of Table 11.2. If the slope, in going from one asymptote to the next, changes by +40 dB/decade there is a pair of complex zeros and the transfer function has a term of the form

$$\frac{s^2 + 2\zeta\omega_n s + \omega_n^2}{\omega_n^2}$$

where ω_n is the frequency at the slope change. Table 11.2 can be used to give an estimate of the damping factor ζ. All the constituent elements of the transfer function are then combined to give a transfer function which has then a gain constant K included. The magnitude is then calculated from the transfer function at some frequency and compared with the experimental value at that frequency, hence K is determined (see Example 10).

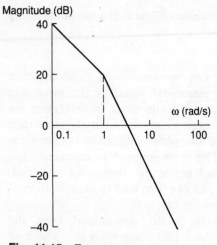

Fig. 11.15 Example 10

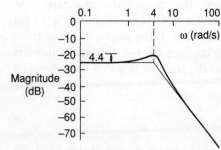

Fig. 11.16 Example 11

Example 10

Determine the transfer function of the system giving the Bode diagram shown in Fig. 11.15. Only the asymptotes are shown.

Answer

There is an initial slope of $-20\,\text{dB/decade}$ and so there is a $1/s$ term. There is a change in slope of $-20\,\text{dB/decade}$ at $\omega = 1\,\text{rad/s}$ and so there is a $1/(\tau s + 1)$ term with $\tau = 1\,\text{s}$. There are no other changes in slope and so the transfer function is of the form

$$G(s) = \frac{K}{s(s + 1)}$$

where K is the gain. The frequency-response function for the system is

$$G(j\omega) = \frac{K}{j\omega(j\omega + 1)} = \frac{K(-\omega^2 - j\omega)}{(-\omega^2 + j\omega)(-\omega^2 - j\omega)} = \frac{-K\omega^2 - jK\omega}{\omega^4 + \omega^2}$$

Hence

$$|G(j\omega)| = \frac{K}{\sqrt{(\omega^4 + \omega^2)}}$$

Figure 11.15 indicates that when $20\lg|G(j\omega)| = 20$ then $\omega = 1\,\text{rad/s}$. Thus

$$20\lg\left[\frac{K}{\sqrt{(1 + 1)}}\right] = 20$$

$$\frac{K}{\sqrt{2}} = 10^1$$

Hence $K = 14.1$. Thus the transfer function is

$$G(s) = \frac{14.1}{s(s + 1)}$$

Example 11

Determine the transfer function of the system giving the Bode diagram shown in Fig. 11.16.

Answer

The initial slope is zero so there is no $1/s$ factor. There is a change in slope of the asymptote of $-40\,\text{dB/decade}$ at about $4\,\text{rad/s}$. The transfer function thus includes the term

$$\frac{\omega_n^2}{s^2 + 2\zeta\omega_n s + \omega_n^2}$$

where $\omega_n = 4\,\text{rad/s}$. The peak is $4.4\,\text{dB}$ above the asymptotes at the break point. Table 11.2 indicates that this corresponds to a damping factor of 0.3. Hence the transfer function has the term

$$\frac{16}{s^2 + 2.4s + 16}$$

There are no other changes in slope and so the transfer function is of the form

$$G(s) = \frac{16K}{s^2 + 2.4s + 16}$$

The frequency-response function is thus

$$G(j\omega) = \frac{16K}{-\omega^2 + j2.4\omega + 16} = \frac{16K[(16 - \omega^2) - j2.4\omega]}{(16 - \omega^2)^2 + 2.4^2\omega^2}$$

Hence the magnitude is

$$|G(j\omega)| = \frac{16K}{\sqrt{[(16 - \omega^2)^2 + 2.4^2\omega^2]}}$$

When ω tends towards 0 then the experimental data gives the magnitude as -25 dB. Thus

$$20\lg(16K/16) = -25$$

$K = 0.056$. The transfer function is thus

$$G(s) = \frac{0.9}{s^2 + 2.4s + 16}$$

Compensation design

Compensation is the term used to describe the adjustment of the performance of a controller in order that it gives a better performance (see Chapter 10). Bode plots can be used to see the effects of changes in compensator design on the performance.

Introducing just a proportional element, i.e. an element with a constant gain, just shifts the magnitude plot up or down and has no effect on the phase plot. A *cascade lead compensator* has a transfer function of the form (equation [25], Chapter 10)

$$G(s) = \frac{K(s + z)}{(s + p)} \tag{41}$$

where $p > z$. This equation can be rearranged to give

$$G(s) = \frac{K(z/p)(1 + s/z)}{(1 + s/p)} \tag{42}$$

The K and the (z/p) terms are constant gain terms, the $(1 + s/z)$ is a real zero term with a time constant of $1/z$, and the $(1 + s/p)$ is a real pole term with a time constant of $1/p$. Since $p > z$ then $(1/z) > (1/p)$. The Bode plot is thus of the form shown in Fig. 11.17. The values given are for $K = 1$ and different ratios of (z/p). When $z = p$ and the pole and zero terms cancel to give $G(s) = 1$, then the magnitude is a straight line along the 0 dB axis and the phase a straight line along the 0° axis. The effect of introducing the lead compensator is thus to lower the magnitude plot at low frequencies and raise the

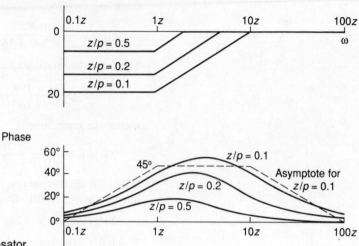

Magnitude (dB)

Fig. 11.17 Bode plot for a lead compensator

overall phase angle, i.e. increase the phase angle lead the output has over the input.

A *cascade lag compensator* has a transfer function of the form (equation [24], Chapter 10)

$$G(s) = \frac{K(s + z)}{(s + p)} \qquad [43]$$

where $z > p$. This equation can be rearranged to give

$$G(s) = \frac{K(z/p)(1 + s/z)}{(1 + s/p)} \qquad [44]$$

The K and the (z/p) terms are constant gain terms, the $(1 + s/z)$ is a real zero term with a time constant of $1/z$, and the $(1 + s/p)$ is a real pole term with a time constant of $1/p$. Since $z > p$ then $(1/p) > (1/z)$. The Bode plot is thus of the form shown in Fig. 11.18. The values given are for $K = 1$ and different ratios of (z/p). When $z = p$ and the pole and zero terms cancel to give $G(s) = 1$ then the magnitude is a straight line along the 0 dB axis and the phase is a straight line along the 0° axis. The effect of introducing the lag compensator is thus to lower the magnitude plot at high frequencies and increase the phase angle lag of the output relative to the input.

Nyquist diagrams

To specify the behaviour of a system to a sinusoidal input (i.e. specify the frequency-response function $G(j\omega)$) at a particular angular frequency ω both the magnitude $|G(j\omega)|$ and the phase ϕ have to be stated. Both the magnitude and phase are

Magnitude (dB)

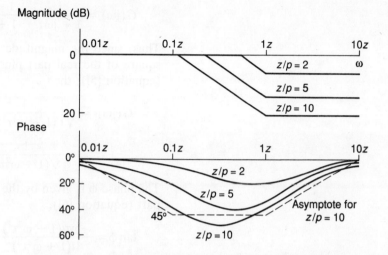

Phase

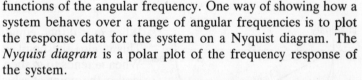

Fig. 11.18 Bode plot for a lag compensator

functions of the angular frequency. One way of showing how a system behaves over a range of angular frequencies is to plot the response data for the system on a Nyquist diagram. The *Nyquist diagram* is a polar plot of the frequency response of the system.

A complex number can be represented by $x + jy$, where x is the real part and y the imaginary part. The number can be plotted as a point on an Argand diagram, i.e. a diagram having the y-axis representing the imaginary part and the x-axis the real part (as in Fig. 11.1). The polar version of the complex number has a line representing $\sqrt{(x^2 + y^2)}$ drawn from the origin at an angle to the axis of ϕ, where $\tan \phi = y/x$. Both methods end up specifying the same point on the Argand diagram.

On the Nyquist diagram the output, for a unit-amplitude sinusoidal input at a particular angular frequency, is specified by drawing a line of length equal to the magnitude $|G(j\omega)|$ at the phase angle ϕ with the real axis, as in Fig. 11.19. The sinusoidal input to the system is thus effectively represented by a line of magnitude 1 lying along the real axis.

In plotting Nyquist diagrams there are four key points to be represented on the diagrams: the start of the plot where $\omega = 0$; the end of the plot where $\omega = \infty$; where the plot crosses the real axis, i.e. $\phi = 0°$ or $\pm180°$; and where it crosses the imaginary axes, i.e. $\phi = \pm90°$. For a *first-order system*, or simple lag system, the transfer function is of the form

$$G(s) = \frac{1}{1 + \tau s} \tag{45}$$

where τ is the time constant. Thus the frequency-response function $G(j\omega)$ is

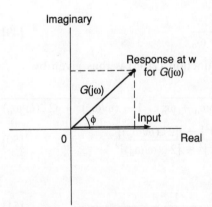

Fig. 11.19 Plotting a point on a Nyquist diagram

$$G(j\omega) = \frac{1}{1 + j\omega\tau} = \frac{1 - j\omega\tau}{1 + \omega^2\tau^2} \qquad [46]$$

Thus, since the magnitude $|G(j\omega)|$ is the square root of the square of the real part plus the square of the imaginary part (equation [5]), then

$$|G(j\omega)| = \sqrt{\left[\frac{1}{(1 + \omega^2\tau^2)^2} + \frac{\omega^2\tau^2}{(1 + \omega^2\tau^2)^2}\right]}$$

$$= \frac{1}{\sqrt{(1 + \omega^2\tau^2)}} \qquad [47]$$

The phase ϕ is given by the imaginary part divided by the real part (equation [6]),

$$\tan\phi = \frac{-\omega\tau/(1 + \omega^2\tau^2)}{1/(1 + \omega^2\tau^2)}$$

$$\phi = -\tan^{-1}\omega\tau \qquad [48]$$

When $\omega = 0$ then equation [48] gives $|G(j\omega)| = 1$ and $\phi = 0°$. This is also the point at which the plot crosses the real axis. When ω tends to ∞ then $|G(j\omega)|$ tends to 0 and ϕ to $-90°$. This is also the point at which the plot crosses the imaginary axis. Other points can be calculated to aid the drawing of the Nyquist plot. Thus, for example, when $\omega = 1/\tau$ then $|G(j\omega)| = 1/\sqrt{2}$ and $\phi = -\tan^{-1}1 = -45°$. Figure 11.20 shows the Nyquist plot. It is a semicircle.

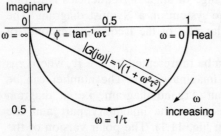

Fig. 11.20 Nyquist plot for $G(s) = 1/(1 + \tau s)$

Now consider a *second-order system* with a transfer function given by

$$G(s) = \frac{\omega_n^2}{s^2 + 2\zeta\omega_n s + \omega_n^2} \qquad [49]$$

The frequency-response function $G(j\omega)$ is thus given by

$$G(j\omega) = \frac{\omega_n^2}{-\omega^2 + j2\zeta\omega\omega_n + \omega_n^2} = \frac{1}{1 - (\omega/\omega_n)^2 + j2\zeta(\omega/\omega_n)}$$

$$G(j\omega) = \frac{[1 - (\omega/\omega_n)^2] - j[2\zeta(\omega/\omega_n)]}{[1 - (\omega/\omega_n)^2]^2 + [2\zeta(\omega/\omega_n)]^2} \qquad [50]$$

and so the magnitude is

$$|G(j\omega)| = \frac{1}{\sqrt{\{[1 - (\omega/\omega_n)^2]^2 + [2\zeta(\omega/\omega_n)]^2\}}} \qquad [51]$$

and the phase is given by

$$\tan\phi = -\frac{2\zeta(\omega/\omega_n)}{1 - (\omega/\omega_n)^2}$$

$$\phi = -\tan^{-1}\left[\frac{2\zeta(\omega/\omega_n)}{1 - (\omega/\omega_n^2)^2}\right] \qquad [52]$$

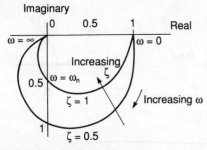

Fig. 11.21 Nyquist plot for
$G(s) = \omega_n^2/(s^2 + 2\zeta\omega_n s + \omega_n^2)$

When $\omega = 0$ then $|G(j\omega)| = 1$ and $\phi = 0°$. When $\omega = \infty$ then $|G(j\omega)| = 0$ and $\phi = -\tan^{-1}(0/-\infty) = -180°$. These are both points at which the plot crosses the real axis. When $\omega = \omega_n$ then $|G(j\omega)| = 1/2\zeta$ and $\phi = -\tan^{-1}(2\zeta/0) = -90°$. This is the point at which the plot crosses the imaginary axis. Figure 11.21 shows the family of Nyquist plots produced for different damping ratios.

Figure 11.22 shows further examples of Nyquist diagrams and their associated root locus diagrams.

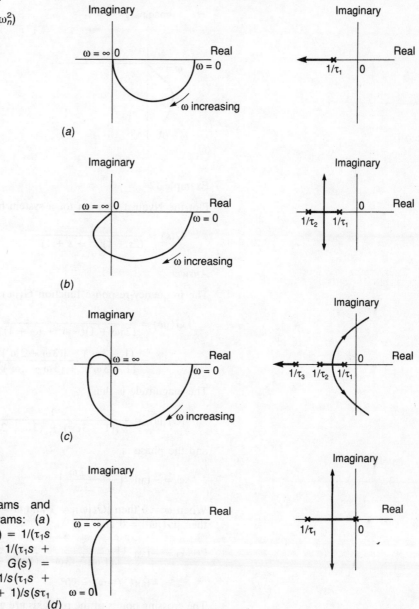

Fig. 11.22 Nyquist diagrams and associated root-locus diagrams: (a) $G(s) = 1/(\tau_1 s + 1)$, (b) $G(s) = 1/(\tau_1 s + 1)(\tau_2 s + 1)$, (c) $G(s) = 1/(\tau_1 s + 1)(\tau_2 s + 1)(\tau_3 s + 1)$, (d) $G(s) = 1/s(\tau_1 s + 1)$, (e) $G(s) = 1/s(\tau_1 s + 1)(\tau_2 s + 1)$, (f) $G(s) = (\tau s + 1)/s(s\tau_1 + 1)(s\tau_2 + 1)$

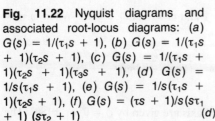

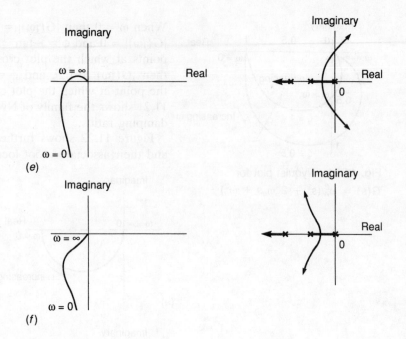

(e)

(f)

Example 12

Plot the Nyquist diagram for a system having a transfer function of

$$G(s) = \frac{1}{(2s + 1)(s^2 + s + 1)}$$

Answer

The frequency-response function $G(j\omega)$ is

$$G(j\omega) = \frac{1}{(2j\omega + 1)(-\omega^2 + j\omega + 1)} = \frac{1}{(1 - 3\omega^2) + j(3\omega - 2\omega^3)}$$

$$= \frac{(1 - 3\omega^2) - j(3\omega - 2\omega^3)}{(1 - 3\omega^2)^2 + (3\omega - 2\omega^3)^2}$$

The magnitude is thus

$$|G(j\omega)| = \frac{1}{\sqrt{[(1 - 3\omega^2)^2 + (3\omega - 2\omega^3)^2]}}$$

and the phase is

$$\phi = -\tan^{-1}\left(\frac{3\omega - 2\omega^3}{1 - 3\omega^2}\right)$$

When $\omega = 0$ then $|G(j\omega)| = 1$ and $\phi = -\tan^{-1}0 = 0°$. When $\omega = \infty$ then $|G(j\omega)| = 0$ and

$$\phi = -\tan^{-1}\left[\frac{3}{(1/\omega) - 3\omega} - \frac{2}{(1/\omega^3) - (3/\omega)}\right]$$

$$= -\tan^{-1}(-\infty) = 90°$$

The crossing points of the real axis are given by $\phi = 0°$ or $\pm 180°$. The

$0°$ point is given by $\omega = 0$, as indicated above. For $\phi = -180°$ then $\tan\phi = -0$ and so we must have

$$3\omega - 2\omega^3 = 0$$

This means $\omega = 0$ or $\sqrt{(3/2)} = 1.2\,\text{rad/s}$. With this value the magnitude has a value of

$$\frac{1}{\sqrt{[(1 - 3\omega^2)^2 + (3\omega - 2\omega^3)^2]}} = \frac{1}{\sqrt{[(1 - 3\{3/2\})^2 + 0]}}$$

or -0.3. The crossing points of the imaginary axis are given by $\phi = \pm 90°$. The $\omega = \infty$ gives the $+90°$ point, as indicated above. For $\phi = -90°$ then $\tan\phi = -\infty$ and so we must have

$$1 - 3\omega^2 = 0$$

This means $\omega = 1/\sqrt{3} = 0.6\,\text{rad/s}$. With this value the magnitude has a value of

$$\frac{1}{\sqrt{[(1 - 3\omega^2)^2 + (3\omega - 2\omega^3)^2]}} = \frac{1}{\sqrt{[0 + (\sqrt{3} - 2/3\sqrt{3})^2]}}$$

or -0.7. To aid in plotting the Nyquist diagram other values can be obtained. Thus for $\omega = 1\,\text{rad/s}$ then $|G(j\omega)| = 0.4$ and $\phi = -153°$, for $\omega = 0.5\,\text{rad/s}$ then $|G(j\omega)| = 0.8$ and $\phi = 68°$, for $\omega = 0.2\,\text{rad/s}$ then $|G(j\omega)| = 0.95$ and $\phi = 34°$. Figure 11.23 shows the resulting Nyquist plot.

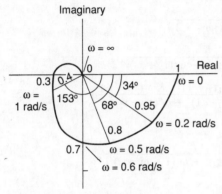

Fig. 11.23 Example 12

Nyquist stability criterion

When there is a sinusoidal input to a system the output from that system is sinusoidal with the same angular frequency but can have an amplitude which differs from that of the input and show a phase difference. The ratio of the output to input amplitudes is the magnitude $|G(j\omega)|$. The basic *Nyquist stability criterion* can be stated as: for instability to occur when the input to a system is sinusoidal then the open-loop magnitude must be greater than 1 when the open-loop phase lag is 180°. If the system causes a phase change of 180° then the feedback signal will be in phase with the input signal and thus will add to it rather than subtract. If the amplitude is less than that of the input signal then a steady condition can be achieved, but if the amplitude is greater then the signal through the system will continuously build up. If the open-loop system is stable then the closed-loop system will be stable.

Figure 11.24 shows the implication of this stability criterion in relation to the Nyquist diagram for an open-loop system. A phase angle of 180° means a magnitude pointing out along the negative real axis. If the magnitude at this phase must not exceed 1 then the polar plot must not enclose the −1 point on the real axis if the system is to be stable.

Example 13

What is the condition for a system with the following open-loop transfer function to be stable?

$$G_o(s) = \frac{K}{s(\tau_1 s + 1)(\tau_2 s + 1)}$$

Answer

The frequency-response function is

$$G_o(j\omega) = \frac{K}{j\omega(j\omega\tau_1 + 1)(j\omega\tau_2 + 1)}$$

$$= \frac{K}{j\omega(-\omega^2\tau_1\tau_2 + 1) - \omega^2(\tau_1 + \tau_2)}$$

$$= \frac{-jK\omega(-\omega^2\tau_1\tau_2 + 1) - K\omega^2(\tau_1 + \tau_2)}{\omega^2(-\omega^2\tau_1\tau_2 + 1)^2 + \omega^4(\tau_1 + \tau_2)^2}$$

The magnitude is thus

$$|G_o(j\omega)| = \frac{K}{\sqrt{[\omega^4(\tau_1 + \tau_2)^2 + \omega^2(1 - \omega^2\tau_1\tau_2)^2]}}$$

and the phase

$$\phi = \tan^{-1}\left[\frac{1 - \omega^2\tau_1\tau_2}{\omega(\tau_1 + \tau_2)}\right]$$

For stability the magnitude must not exceed 1 when the phase is 180°. Thus the limiting condition for the phase is $\phi = \tan^{-1}0$. Hence

$$1 - \omega^2\tau_1\tau_2 = 0$$

and so $\omega = 1/\sqrt{(\tau_1\tau_2)}$. Substituting this value of ω into the magnitude equation gives

$$|G_o(j\omega)| = \frac{K}{\sqrt{[(1/\tau_1\tau_2)^2(\tau_1 + \tau_2)^2 + (1/\tau_1\tau_2)(1 - 1)^2]}}$$

If the system is to be stable then

$$\frac{K}{(1/\tau_1\tau_2)(\tau_1 + \tau_2)} < 1$$

$$K < \frac{\tau_1 + \tau_2}{\tau_1\tau_2}$$

The Nyquist plot is of the form shown in Fig. 11.22(*e*).

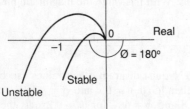

Fig. 11.24 Stable and unstable Nyquist plots

Gain margin and phase margin

The *gain margin* is defined as being the factor by which the system gain, i.e. magnitude, can be increased before instability occurs. It is thus the amount by which the magnitude at 180° has to be increased to reach the critical value of 1 (Fig. 11.25).

$$1 = \text{Gain margin} \times |G(j\omega)|_{\phi = 180°} \qquad [53]$$

It is normally quoted in decibels and so in decibels is

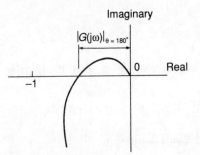

Fig. 11.25 Gain margin

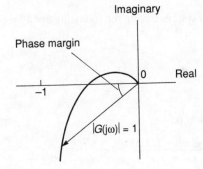

Fig. 11.26 Phase margin

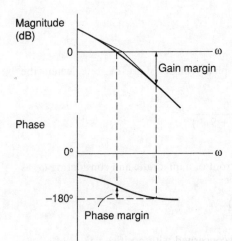

Fig. 11.27 Gain and phase margins with Bode plots

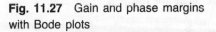

$$\text{Gain margin} = 20 \lg 1 - 20 \lg |G(j\omega)|_{\phi=180°}$$

$$\text{Gain margin} = -20 \lg |G(j\omega)|_{\phi=180°} \qquad [54]$$

If the Nyquist plot never crosses the negative real axis then the gain margin is infinite. If the plot passes through the axis at a value less than 1 then the gain margin is positive. If it passes through the axis at 1 then the gain margin is zero and if it passes through the axis at a value greater than 1, i.e. the plot encloses the −1 point, then the gain margin is negative.

The *phase margin* is defined as the angle through which the Nyquist plot must be rotated in order that the unity magnitude point on the plot passes through the −1 point on the real axis (Fig.11.26). It is thus the amount by which the open-loop phase falls short of 180° when the magnitude has the value 1, i.e. the output amplitude is the same as that of the input.

Figure 11.27 shows the gain margin and phase margin on the Bode plot for a system.

Example 14

Determine the value of K for a system with the following open-loop transfer function

$$G_o(s) = \frac{K}{s(2s + 1)(s + 1)}$$

which will give (*a*) a marginally stable system, (*b*) a gain margin of 3 dB.

Answer

The open-loop frequency response function is

$$G_o(j\omega) = \frac{K}{j\omega(j2\omega + 1)(j\omega + 1)} = \frac{K}{-3\omega^2 + j\omega(1 - 2\omega^2)}$$

$$= \frac{-3K\omega^2 - jK\omega(1 - 2\omega^2)}{9\omega^4 + \omega^2(1 - 2\omega^2)^2}$$

The magnitude is thus

$$|G_o(j\omega)| = \frac{K}{\sqrt{[9\omega^4 + \omega^2(1 - 2\omega^2)^2]}}$$

and the phase

$$\phi = \tan^{-1}\left(\frac{1 - 2\omega^2}{3\omega}\right)$$

(*a*) For the system to be marginally stable the magnitude must have the value 1 when $\phi = 180°$. For $\phi = 180°$ then $\phi = \tan^{-1} 0$ and so

$$1 - 2\omega^2 = 0$$

Hence $\omega = 1/\sqrt{2}$ rad/s. Substituting this value in the magnitude equation, with the magnitude equal to 1, gives

$$1 = \frac{K}{\sqrt{(9/4 + 0)}}$$

and so $K = 1.5$

(b) For the system to have a gain margin of $3\,dB$ then, using equation [54],

$$\text{Gain margin} = -20 \lg |G(j\omega)|_{\phi = 180°}$$

$$3 = -20 \lg \left[\frac{K}{\sqrt{(9/4 + 0)}} \right]$$

and so $K = 1.06$.

Example 15

What is the phase margin for a system having the following open-loop transfer function?

$$G_o(s) = \frac{9}{s(s + 3)}$$

Answer

The frequency-response function is

$$G_o(j\omega) = \frac{9}{-\omega^2 + j3\omega} = \frac{-9\omega^2 - j27\omega}{\omega^4 + 9\omega^2}$$

Hence the magnitude is

$$|G_o(j\omega)| = \frac{9}{\sqrt{(\omega^4 + 9\omega^2)}}$$

and the phase

$$\phi = \tan^{-1}\left(\frac{3}{\omega}\right)$$

The phase margin is the phase angle short of $180°$ when the magnitude is 1. Thus

$$1 = \frac{9}{\sqrt{(\omega^4 + 9\omega^2)}}$$

$$\omega^4 + 9\omega^2 - 81 = 0$$

Using the expression for the root of a quadratic and considering ω^2 as the root, then

$$\omega^2 = \frac{-9 \pm \sqrt{(81 + 324)}}{2}$$

and hence since we are only concerned with positive values of ω, we have $\omega = 2.36\,rad/s$. The phase angle for this magnitude is thus

$$\phi = \tan^{-1}\left(\frac{3}{2.36}\right) = 51.8°$$

Since both the real and the imaginary parts of $G_o(j\omega)$ are negative, then this angle is the angle relative to $-180°$ and so is the phase margin.

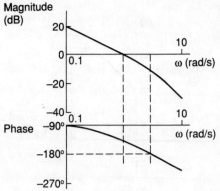

Example 16

For the Bode plot given in Fig. 11.28, estimate the gain margin and the phase margin.

Answer

The gain margin is the magnitude value when the phase is $-180°$ and so is about $8\,\text{dB}$. The phase margin is the phase difference from $-180°$ when the magnitude is zero. It is thus about $40°$.

Fig. 11.28 Example 16

Problems

1 What are the magnitudes and phases of the systems having the following transfer functions?

(a) $\dfrac{5}{s + 2}$

(b) $\dfrac{2}{s(s + 1)}$

(c) $\dfrac{1}{(2s + 1)(s^2 + s + 1)}$

2 What will be the steady-state response of a system with a transfer function

$$\frac{1}{s + 2}$$

when subject to the input?

$$\theta_i = 3\sin(5t + 30°)$$

3 What will be the steady-state response of a system with a transfer function

$$\frac{5}{s^2 + 3s + 10}$$

when subject to the input?

$$\theta_i = 2\sin(2t + 70°)$$

4 What are the magnitude and the gain for a system giving the transfer function?

$$G(s) = \frac{10}{s(s + 1)(s + 2)}$$

5 For systems with the following transfer functions give the values of the magnitude and phase at angular frequencies of (i) $0\,\text{rad/s}$, (ii) $1\,\text{rad/s}$, (iii) $2\,\text{rad/s}$, (iv) $\infty\,\text{rad/s}$.

(a) $G(s) = \dfrac{1}{s(2s + 1)}$

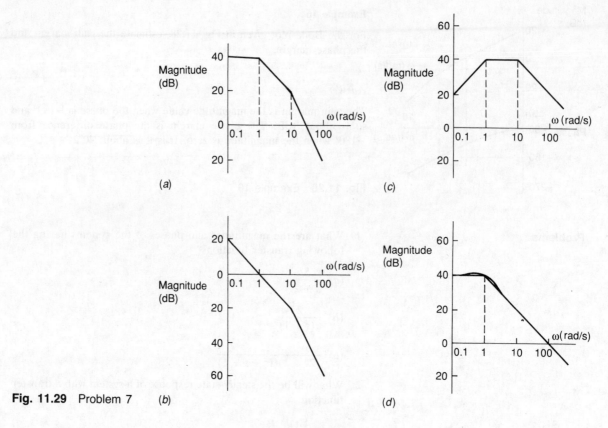

Fig. 11.29 Problem 7   (b)

(b) $G(s) = \dfrac{1}{3s + 1}$

6 Draw the asymptotes of the Bode plots for the systems having the transfer functions

(a) $G(s) = \dfrac{10}{s(0.1s + 1)}$

(b) $G(s) = \dfrac{1}{(2s + 1)(0.5s + 1)}$

(c) $G(s) = \dfrac{5}{(2s + 1)(s^2 + 3s + 25)}$

(d) $G(s) = \dfrac{(s + 1)}{(s + 2)(s + 3)}$

7 Determine the transfer functions of the systems giving the Bode magnitude plot asymptotes shown in Fig. 11.29.

8 Plot the Nyquist diagram for the system having the transfer function

$$G(s) = \dfrac{1}{s(s + 1)}$$

9 Determine the condition for stability for systems having open-loop transfer functions of

(a) $G_o(s) = \dfrac{K}{s^2(\tau s + 1)}$

(b) $G_o(s) = \dfrac{K}{(\tau_1 s + 1)(\tau_2 s + 1)(\tau_3 s + 1)}$

(c) $G_o(s) = \dfrac{K}{s(\tau s + 1)}$

10 Determine the value of K for a system with the following open-loop transfer function

$$G_o(s) = \frac{K}{s(2s + 1)(3s + 1)}$$

which will give (a) a marginally stable system, (b) a gain margin of 2dB.

11 What is the phase margin for a system having the open-loop transfer function?

$$G_o(s) = \frac{2}{s(s + 1)}$$

12 For the Bode plot given in Fig. 11.30, estimate the gain margin and the phase margin.

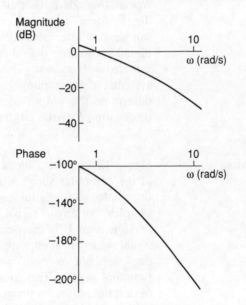

Fig. 11.30 Problem 12

12 Digital control

A digital computer used as a controller in a feedback control system has the advantage over the analogue form of controller in that changes to the control law of an analogue controller require changes to the hardware while changes to the control law of a digital computer can often be produced just by changes in software, i.e. changes in a line or lines of the computer program. A digital computer requires an input of signals in digital form. However the plant being controlled has signals that vary continuously with time, i.e. are analogue signals. Signals have thus to be converted from analogue to digital form. Such a conversion requires the signal to be sampled at intervals, each sample then being converted into digital form. The digital controller can thus be described as a *sampled-data system*. This chapter is a consideration of such systems, the sampling process, the use of z-transforms to determine the output of such systems, and stability. A brief discussion of digital control systems was given in Chapter 1.

The digital control system

Figure 12.1 shows a block diagram of such a system. A clock in the computer supplies a pulse every T seconds. Each time the analogue-to-digital converter (ADC) receives a pulse it samples the error signal. The error signal, a continuously variable signal, is then converted by the ADC into a digital signal which is then supplied to the computer. The error is thus sampled every T seconds, T being referred to as the *sampling period*. The computer then applies the control law, as determined by its program, and its output is then converted into an analogue signal by the digital-to-analogue converted (DAC). The resulting analogue signal can then be used to actuate the correction unit and so control the plant variable.

To indicate which signals are varying continuously with time and which are sampled, a starred symbol is used for all sampled signals. Thus the error signal prior to the ADC is

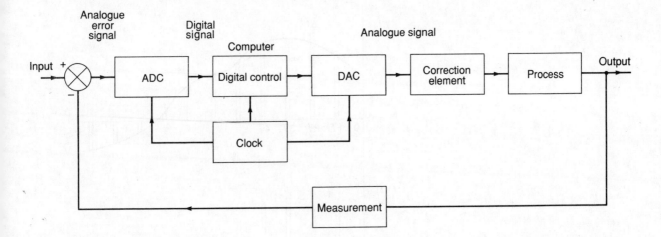

Fig. 12.1 Digital control system with output sampling

represented by $e(t)$, indicating it is a continuous function of time, while after the ADC when the signal has been sampled and converted to digital form the symbol becomes $e^*(t)$.

The sampling process

Fig. 12.2 The analogue-to-digital converter

The ADC samples analogue signals. Generally, sampling occurs at regular time intervals, every T seconds. The sampling element senses the value of the signal over a short interval of time Δt in each sampling period and ignores it for the rest of the period. The ADC can thus be considered to be a switch which is switched on every T seconds for just Δt (Fig. 12.2). Figure 12.3 illustrates this sampling process for a continuous time signal $f(t)$. Δt is small enough for each sampled signal to be of constant size during that time. The output is thus a series of pulses at regular time intervals, the height of a pulse at 0, $1T$, $2T$, $3T$, $4T$, . . ., kT, being a measure of the size of the continuous time signal $f(t)$ at that time. We can represent the series of pulses up to some time kT as

$$f(t = 0), f(t = 1T), f(t = 2T), f(t = 3T), \ldots, f(t = kT),$$

where k is an integer and $f(t = 0)$ is the value of the function at $t = 0$, $f(t = 1T)$ is the value at $t = 1T$, etc.

For a continuous time function $f(t)$ which is a unit step at $t = 0$ (Fig. 12.4), then the series of pulses produced is

$$1, 1, 1, 1, \ldots, 1$$

For a continuous time function $f(t)$ which is a ramp with a slope which increases by 1 for each sampling period (Fig. 12.5) the series of pulses produced is

$$0, 1, 2, 3, \ldots, k$$

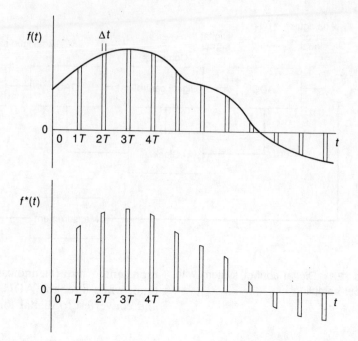

Fig. 12.3 The sampling process

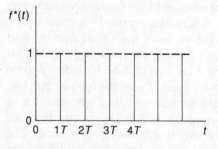

Fig. 12.4 Unit step

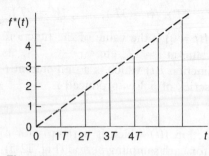

Fig. 12.5 Ramp

For a unit impulse at $t = 0$ (Fig. 12.6(a)) the series of pulses produced is

$$1, 0, 0, 0 \ldots, 0$$

For a unit impulse at $t = 2T$ (Fig. 12.6(b)) the series of pulses produced is

$$0, 0, 1, 0, \ldots, 0$$

The unit step in Fig. 12.4 can be considered as the sum of a series of pulses produced by unit impulses at $t = 0$, $t = 1T$, $t = 2t$, $t = 3T$, ..., $t = kT$. The ramp in Fig. 12.5 can be considered as the sum of a series of pulses produced by impulses of size 0 at $t = 0$, size 1 at $t = 1T$, size 2 at $t = 2T$, size 3 at $t = 3T$, ..., size k at $t = kT$.

Whatever the form of the continuous time function the digital output is a sequence of impulses. Each impulse in the sequence is a unit impulse multiplied by the value of $f(t)$ at that time. We can thus write for a function $f^*(t)$ which describes the sequence of pulses for a function $f(t)$ with sampling period T

$$f^*(t) = f(0)(\text{impulse at } 0) + f(1T)(\text{impulse at } 1T)$$
$$+ f(2T)(\text{impulse at } 2T)$$
$$+ \ldots f(kT)(\text{impulse at } kT) \quad [1]$$

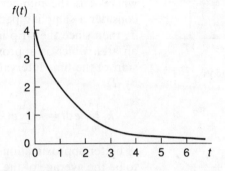

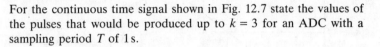

Fig. 12.6 Unit impulses at (a) $t = 0$, (b) $t = 2T$

Example 1

For the continuous time signal shown in Fig. 12.7 state the values of the pulses that would be produced up to $k = 3$ for an ADC with a sampling period T of 1 s.

Fig. 12.7 Example 1

Answer

At $k = 0$ (i.e. $t = 0$), then $f(t) = 4$. At $k = 1$ (i.e. $t = 1T$), then $f(t) = 2$. At $k = 2$ (i.e. $t = 2T$), then $f(t) = 1$. At $k = 3$ (i.e. $t = 3T$), then $f(t) = 0.5$. Thus the digital signal is a series of impulses:

size 4 at $k = 0$, or $t = 0$
size 2 at $k = 1$, or $t = 1$
size 1 at $k = 2$, or $t = 2$
size 0.5 at $k = 3$, or $t = 3$

Digital control laws

Digital controllers can be used to implement the types of control laws discussed for analogue controllers in Chapter 10, i.e. proportional, integral and derivative. The PID control law has the form (equation [15], Chapter 10) that results in an output from the controller of

$$\text{Output} = K_p e + K_i \int_0^t e \, dt + K_d \frac{de}{dt} \qquad [2]$$

where e is the error input to the controller, K_p the proportional constant, K_d the derivative constant and K_i the integral constant. The transfer function $G_c(s)$ of the controller is thus (equation [16], Chapter 10)

$$G_c(s) = K_p + \frac{K_i}{s} + K_d s \qquad [3]$$

The integral term in equation [2] represents the area under a graph of error against time between time $t = 0$ and t. It can be given approximately by dividing the area into rectangular

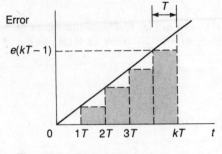

Fig. 12.8 Digital integration

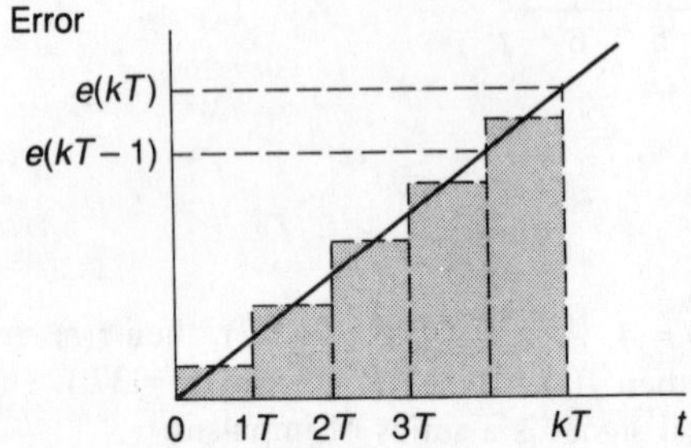

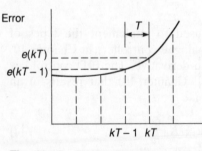

Fig. 12.9 Digital integration

Error

e(kT)

e(kT − 1)

kT − 1 kT   t

Fig. 12.10 Digital differentiation

strips (Fig. 12.8) and then summing the areas of the strips. Thus

$$k_i \int_0^t e \, dt = k_i \sum_0^k \text{ of areas of strips} \qquad [4]$$

where k is the number of strips between $t = 0$ and t. If we consider a sampling period of T and each strip to be of width T, then since the strip immediately preceding the time kT has an area which is approximately the value of the error at the start of the time interval of the strip, i.e. $e(kT − 1)$, multiplied by T

$$k_i \int_0^{kT} e \, dt = k_i \sum_0^k e(kT − 1)T$$

A better approximation is given by taking the height of a strip to be the average of the error values at the start and end of the strip (Fig. 12.9), i.e. the values $e(kT − 1)$ and $e(kT)$. Then

$$k_i \int_0^{kT} e \, dt = k_i \sum_0^k \tfrac{1}{2}[e(kT − 1) − e(kT)]T \qquad [5]$$

Both these equations can be applied to a series of pulses, sampling time T, to arrive at an output which is the integral of the error input.

The derivative term in equation [2] can be considered to be the slope of the error time graph at the time concerned

$$K_d \frac{de}{dt} = K_d(\text{slope of error/time graph}) \qquad [6]$$

For a signal which is a series of pulses, then a reasonable approximation to the slope of the error/time graph is given by the slope of the line joining two consecutive pulses (Fig. 12.10), i.e. pulses at kT and $(kT − 1)$. If these pulses have sizes of $e(kT)$ and $e(kT − 1)$ and the time between the pulses is the sampling period T, then equation [6] can be written as

$$K_d \frac{de}{dt} = \frac{k_o}{T}[e(kT) − e(kT − 1)] \qquad [7]$$

Equation [2] for PID control can thus be written, using equations [5] and [7], as

$$\text{Output} = K_d e(kT) + k_i \sum_0^k \tfrac{1}{2}[e(kT − 1) − e(kT)]T$$

$$+ (k_d/T)[e(kT) − e(kT − 1)] \qquad [8]$$

Digital-to-analogue converter

The output from the digital controller is a binary-coded digital signal. This usually has to be converted into an analogue signal

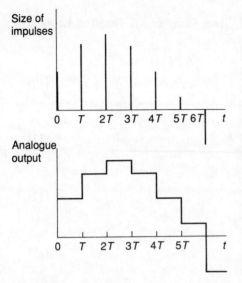

Fig. 12.11 Analogue output from DAC with zero-order hold

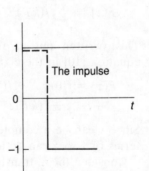

Fig. 12.12 Unit impulse response of a zero-order hold system

to actuate the correction unit. The DAC converts the binary-coded signal, at some instant, into an impulse whose size is related to the binary-code. Thus the input, which was a sequence of binary coded signals, becomes a sequence of impulses with the time between successive impulses being the sampling time T. With what is termed the *zero-order hold* system the sequence of impulses is converted to an analogue signal by holding the value of an impulse until the next one arrives. The result is a staircase form of analogue signal (Fig. 12.11).

With zero-order hold a unit impulse as input gives rise to a pulse of unit size and width T, as illustrated in Fig. 12.12. This output can be considered to be the sum of two step signals where one starts at $t = 0$ and has a positive height and the other starts at $t = T$ and has a negative height (Fig. 12.13). Since the Laplace transform of a unit step is $1/s$ and a delayed start of T means multiplying the transform by e^{-Ts} (see Chapter 4), then the output is

$$\frac{1}{s} - \frac{e^{-Ts}}{s} = \frac{1 - e^{-Ts}}{s}$$

Since the input is a unit impulse, Laplace transform 1, then the transfer function $G(s)$ for the zero-order system is

$$G(s) = \frac{1 - e^{-Ts}}{s} \qquad [9]$$

Fig. 12.13 A pulse as a sum of two steps

The z-transform

The function describing a sequence of impulses can be written as (equation [1])

$$f^*(t) = f(0)(\text{impulse at } 0) + f(1T)(\text{impulse at } 1T)$$
$$+ f(2T)(\text{impulse at } 2T) + \ldots f(kT)(\text{impulse at } kT)$$

The Laplace transform of an impulse at $t = 0$ is 1. The Laplace transform of an impulse at time T is e^{-Ts}, at time $2T$ it is

e^{-2Ts}, and time kT is e^{-kTs} (see Chapter 4). Thus the Laplace transform of $f^*(t)$ is

$$F^*(s) = f(0)1 + f(T)e^{-Ts} + f(2T)e^{-2Ts}$$
$$+ \ldots f(kT)e^{-kTs} \tag{10}$$

We can represent this as

$$F^*(s) = \sum_{k=0}^{k=\infty} f(kT)e^{-kTs} \tag{11}$$

This can be simplified if we let

$$z = e^{Ts} \tag{12}$$

This can also be written as

$$s = \frac{1}{T}\ln z \tag{13}$$

Using equations [12] and [13] we can write equation [11] in terms of z.

$$F^*(s)_{s=(1/T)\ln z} = \sum_{k=0}^{k=\infty} f(kT)z^{-k}$$

$F^*(s)$, with $s = (1/T)\ln z$, is called the z *transform* and is represented by $F(z)$. Thus

$$F(z) = F^*(s)_{s=(1/T)\ln z} \tag{14}$$

$$F(z) = \sum_{k=0}^{k=\infty} f(kT)z^{-k} \tag{15}$$

This, written out as the sequence of impulse terms, i.e. equation [10] in the z transform, is

$$F(z) = f(0)z^{-0} + f(T)z^{-1} + f(2T)z^{-2}$$
$$+ \ldots + f(kT)z^{-k} \tag{16}$$

Since s can be a complex quantity; then since z is defined in terms of s so z can be a complex quantity.

Consider the z transform for a unit step. Such a step has $f(t) = 1$ for all values of $t > 0$. Thus, using equation [16],

$$F(z) = 1z^{-0} + 1z^{-1} + 2z^{-2} + \ldots + 1z^{-k}$$

For a series of this form, a geometric series, we have the basic relationship

$$1 + x + x^2 + x^3 + \ldots = \frac{1}{1 - x}$$

and thus if $x = 1/z$ then the sum of all these terms in z is

$$F(z) = \frac{1}{1 - 1/z} = \frac{z}{z - 1} \tag{17}$$

This is a more convenient expression to handle than the series of terms for $f^*(t)$ or $F^*(s)$ for the unit step.

Table 12.1 shows the z-transforms for commonly encountered functions and the following are the basic properties of the z-transform:

1 Multiplication of the time function by a constant results in the multiplication of the z-transform by the same constant.

   $kf(t)$ gives $kF(z)$

Table 12.1 z-transforms

$f(t)$	$F(s)$	$F(z)$
Unit impulse	1	1
Unit impulse delayed by kT	e^{-kTs}	z^{-k}
Unit step	$\dfrac{1}{s}$	$\dfrac{z}{z-1}$
Unit step delayed by kT	$\dfrac{e^{-kTs}}{s}$	$\dfrac{z}{z^k(z-1)}$
t (i.e. unit ramp)	$\dfrac{1}{s^2}$	$\dfrac{Tz}{(z-1)^2}$
e^{-at}	$\dfrac{1}{s+a}$	$\dfrac{z}{z-e^{-aT}}$
$1-e^{-at}$	$\dfrac{a}{s(s+a)}$	$\dfrac{z(1-e^{-aT})}{(z-1)(z-e^{-aT})}$
te^{-at}	$\dfrac{1}{(s+a)^2}$	$\dfrac{Tze^{-aT}}{(z-e^{-aT})^2}$
$e^{-at}-e^{-bt}$	$\dfrac{b-a}{(s+a)(s+b)}$	$\dfrac{(e^{-aT}-e^{-bT})z}{(z-e^{-aT})(z-e^{-bT})}$
$(1-at)e^{-at}$	$\dfrac{s}{(s+a)^2}$	$\dfrac{z[z-e^{-aT}(1+aT)]}{(z-e^{-aT})^2}$
$\sin \omega t$	$\dfrac{\omega}{s^2+\omega^2}$	$\dfrac{z\sin\omega T}{z^2-2z\cos\omega T+1}$
$\cos \omega t$	$\dfrac{s}{s^2+\omega^2}$	$\dfrac{z(z-\cos\omega T)}{z^2-2z\cos\omega T+1}$
$e^{-at}\sin \omega t$	$\dfrac{\omega}{(s+a)^2+\omega^2}$	$\dfrac{ze^{-at}\sin\omega T}{(z-e^{(-a+j\omega)T})(z-e^{(-a-j\omega)T})}$
$e^{-at}\cos \omega t$	$\dfrac{s+a}{(s+a)^2+\omega^2}$	$\dfrac{z(z-e^{-aT}\cos\omega T)}{(z-e^{(-a+j\omega)T})(z-e^{(-a-j\omega)T})}$

2   The addition of two time functions gives the addition of their two separate z-transforms.

$$f_1(t) + f_2(t) \text{ gives } F_1(z) + F_2(z)$$

3   The *final-value theorem* gives the value that would be reached eventually by the time function, i.e. the steady-state value. This theorem is

$$\lim_{t \to \infty} f(t) = \lim_{z \to 1} (z - 1)F(z)$$

and applies subject to the condition that $F(z)$ gives a sequence of terms that converges.

4   The *initial-value theorem* gives the value that the time function has when $t = 0$. This theorem is

$$\lim_{t \to 0} f(t) = \lim_{z \to \infty} F(z)$$

and applies subject to the condition that such a limit exists.

5   If the time function is shifted by a time interval kT then the z-transform of the function is multiplied by z^{-k}

$$z \text{ transform of } f(t - kT) = z^{-k}F(z)$$

Example 2

Find the z-transforms of the following sequences:

(a) $f(0) = 1, f(1T) = 1, f(2T) = 1, f(3T) = 1, \ldots, f(kT) = 1.$

(b) $f(0) = 0, f(1T) = 0, f(2T) = 1, f(3T) = 1$, and all further values are 1.

(c) $f(0) = 0, f(1T) = 2, f(2T) = 5, f(3T) = 1$, and all further values are 0.

Answer

(a) This is a unit step starting at $t = 0$. Hence the z-transform is, using the transform for a step given in Table 12.1,

$$F(z) = \frac{z}{z - 1}$$

(b) This is a unit step starting at $t = 2T$. Hence the z-transform is, using the transform for a delayed step given in Table 12.1,

$$F(z) = \frac{z}{z^2(z - 1)}$$

(c) The digital version of this signal is just a series of impulses, each being of the height indicated by the function values. Since the z-transform for a unit impulse delayed by kT is z^{-k}, then applying this to each of the terms in turn gives

$$F(z) = f(0)z^{-0} + f(T)z^{-1} + f(2T)z^{-2} + \ldots + f(kT)z^{-k}$$
$$F(z) = 0 + 2z^{-1} + 5z^{-2} + z^{-3} + 0 + 0 +, \text{ etc.}$$

Inverse z-transform

The sequence of samples represented by a z-transform, i.e. the inverse z-transform, can be obtained in a number of ways, e.g. by an approach based on *long division* or one based on the use of partial fractions. Here only the long-division approach is considered, the reason being that mathematically it is the simplest to apply. Consider the z-transform

$$F(z) = \frac{2z}{z^2 + z + 1}$$

The procedure is to divide the numerator by the denominator by long division.

$$
\begin{array}{r}
2z^{-1} - 2z^{-2} + 2z^{-4} - 2z^{-5} + \ldots \\
z^2 + z + 1 \overline{)2z} \\
2z + 2 + 2z^{-1} \\
-2 - 2z^{-1} \\
-2 - 2z^{-1} - 2z^{-2} \\
2z^{-2} \\
2z^{-2} + 2z^{-3} + 2z^{-4} \\
-2z^{-3} - 2z^{-4} \\
-2z^{-3} - 2z^{-4} - 2z^{-5} \\
2z^{-5}
\end{array}
$$

Thus we can represent $F(z)$ by a series, i.e.

$$F(z) = 2z^{-1} - 2z^{-2} + 2z^{-4} - 2z^{-5} + \ldots$$

Since the z-transform of a unit impulse delayed by a time kT is z^{-k}, then the sequence of samples represented by $F(z)$ is

$$f(1T) = 2, f(2T) = -2, f(4T) = 2, f(5T) = -2, \ldots$$

Figure 12.14 shows the sequence.

Example 3

Using long division, find the inverse z-transforms in the form of the sample sequence for the following:

(a) $F(z) = \dfrac{5}{z + 1}$

(b) $F(z) = \dfrac{4z}{(z - 1)^2}$

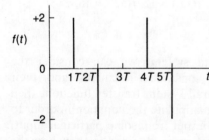

Fig. 12.14 $F(z) = 2z/(z^2 + z + 1)$

Answer

(a) Using long division

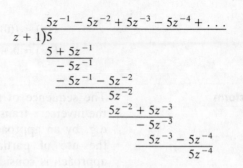

$$
\begin{array}{r}
5z^{-1} - 5z^{-2} + 5z^{-3} - 5z^{-4} + \dots \\
z + 1\overline{)5} \\
5 + 5z^{-1} \\
\hline
- 5z^{-1} \\
- 5z^{-1} - 5z^{-2} \\
\hline
5z^{-2} \\
5z^{-2} + 5z^{-3} \\
\hline
- 5z^{-3} \\
- 5z^{-3} - 5z^{-4} \\
\hline
5z^{-4}
\end{array}
$$

Thus

$$F(z) = 5z^{-1} - 5z^{-2} + 5z^{-3} - 5z^{-4} + \dots$$

Hence the inverse transform is a series of samples

$$f(1T) = 5, \ f(2T) = -5, \ f(3T) = +5, \ f(4T) = -5, \dots$$

(b) Using long division

$$
\begin{array}{r}
4z^{-2} + 8z^{-3} + 12z^{-4} + 16z^{-5} + \dots \\
z^2 - 2z + 1\overline{)4} \\
4 - 8z^{-1} + 4z^{-2} \\
\hline
8z^{-1} - 4z^{-2} \\
8z^{-1} - 16z^{-2} + 8z^{-3} \\
\hline
12z^{-2} - 8z^{-3} \\
12z^{-2} - 24z^{-3} + 12z^{-4} \\
\hline
16z^{-3} - 12z^{-4} \\
16z^{-3} - 32z^{-4} + 16z^{-5} \\
\hline
20z^{-4} - 16z^{-5}
\end{array}
$$

Thus

$$F(z) = 4z^{-2} + 8z^{-3} + 12z^{-4} + 16z^{-5} + \dots$$

Hence the inverse transform is a series of samples

$$f(2T) = 4, \ f(3T) = 8, \ f(4T) = 12, \ f(5T) = 16, \dots$$

Sampled-data systems

In considering sampled-data systems we can consider the transfer functions of the constituent elements and their combination to give an overall system transfer function, then carry out a z-transform to manipulate the equation in order to obtain a description of the output from some particular input. Finally the inverse transform can be obtained to show how the signal varies with time. The following example illustrates this.

Consider the system, an *electrical network*, shown in Fig. 12.15. This has the transfer function

$$G(s) = \frac{1/RC}{s + (1/RC)} = \frac{0.25}{s + 0.25}$$

The z-transform of the transfer function is

$$G(z) = z \text{ transform of } G(s)$$

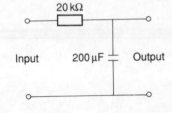

Fig. 12.15 $G(s) = 0.25/(s + 0.25)$

$$= z \text{ transform of } \left(\frac{0.25}{s + 0.25} \right)$$

This expression is of the form $a/(s + a)$ and Table 12.1 gives the z-transform of this as $z/(z - e^{-aT})$, where T is the sampling period. Thus

$$G(z) = \frac{z}{z - e^{-0.25T}} \qquad [18]$$

If we consider a sampling period of 1s then equation [18] becomes

$$G(z) = \frac{z}{z - 0.78} \qquad [19]$$

The transfer function $G(s)$ describes the relationship between the input and the output when both are in the s domain. Likewise, the transfer function $G(z)$ describes the relationship between the input and the output when both are in the z-domain.

$$G(s) = \frac{\text{output}(s)}{\text{input}(s)}$$

$$G(z) = \frac{\text{output}(z)}{\text{input}(z)}$$

Thus if we consider there to be an impulse input, then since the z-transform of an impulse is 1, equation [19] gives the output (z) as

$$\text{Output}(z) = \frac{z}{z - 0.78} \qquad [20]$$

In order to carry out the inverse transformation to see how the output varies with time, the expression is reorganized by means of long division.

$$
\begin{array}{r}
1 + 0.78z^{-1} + 0.61z^{-2} + 0.47z^{-3} + 0.37z^{-4} + \ldots \\
z - 0.78 \overline{)z} \\
\underline{z - 0.78} \\
+ 0.78 \\
\underline{0.78 - 0.61z^{-1}} \\
+ 0.61z^{-1} \\
\underline{0.61z^{-1} - 0.47z^{-2}} \\
0.47z^{-2} \\
\underline{0.47z^{-2} - 0.37z^{-3}} \\
0.37z^{-3}
\end{array}
$$

Thus equation [20] can be written as

$$\text{Output}(z) = 1 + 0.78z^{-1} + 0.61z^{-2} + 0.47z^{-3} + 0.37z^{-4}$$
$$+ \ldots$$

The inverse transform of this is an output which gives the following series of pulses

$$f(0) = 1, f(1) = 0.78, f(2) = 0.61, f(3) = 0.47,$$
$$f(4) = 0.37, \ldots$$

We can determine the output that the system would have when transients have died away by using the final-value theorem

$$\lim_{t \to \infty} f(t) = \lim_{z \to 1} (z - 1)F(z)$$

$F(z)$ is given by equation [20] and thus

$$(z - 1)F(z) = \frac{(z - 1)z}{z - 0.78}$$

As $z \to 1$ then the expression tends to the value 0. Thus

$$\lim_{t \to \infty} f(t) = 0$$

and so the sizes of the pulses drop eventually to reach 0, i.e. the steady-state value is 0. Figure 12.16 shows the form of the output from the system.

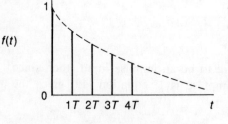

$f(t)$

1T 2T 3T 4T

Fig. 12.16 Response of the system in Fig. 12.15 to an impulse input

Example 4

What is the z-transform of the following transfer function?

$$G(s) = \frac{s}{s + 0.2}$$

Answer

The transform function has to be rearranged to put it into a form which appears in Table 12.1. Thus

$$G(s) = 1 - \frac{0.2}{s + 0.2}$$

The z-transform of $1/(s + a)$ is $z/(z - e^{-aT})$ and for 1 it is 1. Thus

$$G(s) = 1 - 0.2 \left(\frac{z}{z - e^{-0.2T}} \right)$$

$$= \frac{z - e^{-0.2T} - 0.2z}{z - e^{-0.2T}}$$

$$= \frac{0.8z - e^{-0.2T}}{z - e^{-0.2T}}$$

Open-loop sampled-data system

Consider a plant which has a transfer function of

$$G_p(s) = \frac{1}{s(s + 1)}$$

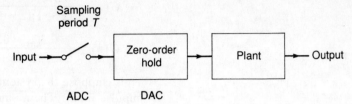

Fig. 12.17 An open-loop sampled-data system

and is used with a zero-order hold system to give an *open-loop sampled-data system* of the form shown in Fig. 12.17. The zero-hold system has a transfer function of (equation [8])

$$G_{zoh}(s) = \frac{1 - e^{-Ts}}{s}$$

Thus the transfer function of the open-loop system is

$$G(s) = G_p(s)G_{zoh}(s)$$

$$= \frac{1 - e^{-Ts}}{s^2(s + 1)} \qquad [21]$$

To obtain the output of such a system for some input we can use the z-transformation.

$$G(z) = z \text{ transform of } G(s)$$

In order to carry out this transform we need to put equation [21] into a form which will enable Table 12.1 to be used. Using partial fractions, equation [21] can be rewritten as

$$G(s) = (1 - e^{-Ts})\left(\frac{1}{s} - \frac{1}{s^2} + \frac{1}{s + 1}\right)$$

Hence

$$G(z) = \left(1 - \frac{1}{z}\right)\left(\frac{Tz}{(z - 1)^2} - \frac{z}{z - 1} + \frac{z}{z - e^{-T}}\right) \qquad [22]$$

Suppose we have a sampling period T of 1 s. Equation [22] then becomes

$$G(z) = \left(1 - \frac{1}{z}\right)\left(\frac{z}{(z - 1)^2} - \frac{z}{z - 1} + \frac{z}{z - e^{-1}}\right)$$

which simplifies to

$$G(z) = \left(\frac{z - 1}{z}\right)$$

$$\left[\frac{z(z - e^{-1}) - z(z - 1)(z - e^{-1}) + z(z - 1)^2}{(z - 1)^2(z - e^{-1})}\right]$$

$$G(z) = \frac{ze^{-1} + 1 - 2e^{-1}}{(z - 1)(z - e^{-1})}$$

$$G(z) = \frac{0.37z + 1 - 0.74}{(z - 1)(z - 0.37)} \qquad [23]$$

$$G(z) = \frac{0.37z + 0.26}{z^2 - 1.37z + 0.37} \qquad [24]$$

Thus, suppose the system described by equation [24] has a unit impulse input. Then, since the z-transform of the input is 1,

$$\text{Output}(z) = \frac{0.37z + 0.26}{z^2 - 1.37z + 0.37} \qquad [25]$$

We can obtain this equation as a series of terms by long division.

$$
\begin{array}{r}
0.37z^{-1} + 0.77z^{-2} + 0.91z^{-3} + 0.97z^{-4} + \ldots \\
\hline
z^2 - 1.37z + 0.37 \overline{)0.37z + 0.26} \\
0.37z - 0.51 + 0.14z^{-1} \\
\hline
+ 0.77 - 0.14z^{-1} \\
0.77 - 1.05z^{-1} + 0.28z^{-2} \\
\hline
+ 0.91z^{-1} - 0.28z^{-2} \\
0.91z^{-1} - 1.25z^{-2} + 0.34z^{-3} \\
\hline
- 0.97z^{-2} - 0.34z^{-3}
\end{array}
$$

Hence

$$\text{Output } (z) = 0.37z^{-1} + 0.77z^{-2} + 0.91z^{-3} + 0.97z^{-4}$$
$$+ \ldots$$

The inverse transform of this is thus an output which varies with time, giving the following signal values:

$$f(0) = 0, \ f(1) = 0.37, \ f(2) = 0.77, \ f(3) = 0.91,$$
$$f(4) = 0.97, \ \ldots$$

We can determine the output that the system would have when transients have died away by using the final-value theorem

$$\lim_{t \to \infty} f(t) = \lim_{z \to 1} (z - 1)F(z)$$

$F(z)$ is given by equation [23], when there is an impulse input, as

$$F(z) = \frac{0.37z + 1 - 0.74}{(z - 1)(z - 0.37)}$$

The equation is used in this form, rather than that in equation [24], since it is to be multiplied by $(z - 1)$. Thus

$$(z - 1)F(z) = \frac{0.37z + 1 - 0.74}{z - 0.37}$$

When $z = 1$ then $(z - 1)F(z) = 1$. Thus at $t \to \infty$ then $f(t) \to 1$. The output is thus of the form shown in Fig. 12.18.

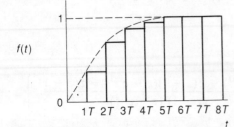

Fig. 12.18 Response of the system in Fig. 12.7 to impulse input

Example 5

Determine the response of the system described by Fig. 12.17 to a unit step input. The sampling period is 1 s.

Answer

The analysis given earlier indicates that the z-transform of the system transfer function is given by equation [23] as

$$G(z) = \frac{0.37z + 1 - 0.74}{(z - 1)(z - 0.37)}$$

A unit step input has the z-transform of $z/(z - 1)$. Thus the output is given by

$$\text{Output}(z) = \frac{0.37z^2 + 0.26z}{(z - 1)^2(z - 0.37)}$$

$$= \frac{0.37z^2 + 0.26z}{z^3 - 2.37z^2 + 1.74z - 0.37}$$

Using long division gives

$$
\begin{array}{r}
0.37z^{-1} + 1.15z^{-2} + 3.31z^{-3} + \dots \\
z^3 - 2.32z^2 + 1.74z - 0.37 \overline{)\, 0.37z^2 + 0.26z} \\
0.37z^2 + 0.89z + 0.64 - 0.14z^{-1} \\
+ 1.15z + 0.64 + 0.14z^{-1} \\
1.15z - 2.67 - 2.00z^{-1} - 0.43z^{-2} \\
+ 3.31 + 2.14z^{-1} + 0.43z^{-2}
\end{array}
$$

We can determine the output that the system would have when transients have died away by using the final-value theorem

$$\lim_{t \to \infty} f(t) = \lim_{z \to 1} (z - 1)F(z)$$

$$(z - 1)F(z) = \frac{(z - 1)(0.37z^2 + 0.26z)}{(z - 1)^2(z - 0.37)}$$

When $z \to 1$ then $(z - 1)F(z) \to \infty$.

Closed-loop sampled-data system

Consider the *closed-loop sampled-data system* shown in Fig. 12.19 with the plant having the transfer function

$$G_p(s) = \frac{1}{s(s + 1)}$$

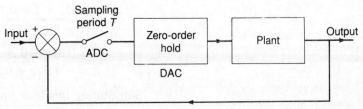

Fig. 12.19 A closed-loop sampled-data system

Since the zero-order hold system has a transfer function of (equation [8])

$$G_{zoh}(s) = \frac{1 - e^{-Ts}}{s}$$

then the transfer function of the closed-loop system is, since there is unity feedback,

$$G(s) = \frac{G_p(s)G_{zoh}(s)}{1 + G_p(s)G_{zoh}(s)}$$

The z-transform of this is

$$G(z) = \frac{G_p(z)G_{zoh}(z)}{1 + G_p(z)G_{zoh}(z)} \tag{26}$$

But

$$G_p(s)G_{zoh}(s) = \frac{1 - e^{-Ts}}{s^2(s + 1)}$$

By means of partial fractions this can be written as

$$G_p(s)G_{zoh}(s) = (1 - e^{-Ts})\left(\frac{1}{s^2} - \frac{1}{s} + \frac{1}{s + 1}\right)$$

Hence

$$G_p(z)G_{zoh} = \left(1 - \frac{1}{z}\right)\left(\frac{Tz}{(z - 1)^2} - \frac{z}{z - 1} + \frac{z}{z - e^{-T}}\right)$$

Suppose we have a sampling period T of 1 s, then

$$G_p(z)G_{zoh} = \left(1 - \frac{1}{z}\right)\left(\frac{z}{(z - 1)^2} - \frac{z}{z - 1} + \frac{z}{z - e^{-1}}\right)$$

which simplifies to

$$G_p(z)G_{zoh}$$

$$= \left(\frac{z - 1}{z}\right)$$

$$\left[\frac{z(z - e^{-1}) - z(z - 1)(z - e^{-1}) + z(z - 1)^2}{(z - 1)^2(z - e^{-1})}\right]$$

$$= \frac{ze^{-1} + 1 - 2e^{-1}}{(z - 1)(z - e^{-1})}$$

$$= \frac{0.37z + 1 - 0.74}{(z - 1)(z - 0.37)}$$

$$= \frac{0.37z + 0.26}{z^2 - 1.37z + 0.37} \tag{27}$$

Substituting this into equation [26] gives

$$G(z) = \frac{0.37z + 0.26}{z^2 - 1.37z + 0.37 + 0.37z + 0.26}$$

$$= \frac{0.37z + 0.26}{z^2 - z + 0.63} \qquad [28]$$

If we now consider there to be a unit impulse input then, since the z-transform of such an input is 1,

$$\text{Output}(z) = \frac{0.37z + 0.26}{z^2 - z + 0.63} \qquad [29]$$

By long division we can obtain

$$
\begin{array}{r}
0.37z^{-1} + 0.63z^{-2} + 0.40z^{-3} - 0.25z^{-5} + \ldots \\
z^2 - z + 0.63 \overline{)0.37z + 0.26} \\
\underline{0.37z - 0.37 + 0.23z^{-1}} \\
+ 0.63 - 0.23z^{-1} \\
\underline{0.63 - 0.63z^{-1} + 0.40z^{-2}} \\
+ 0.40z^{-1} - 0.40z^{-2} \\
\underline{0.40z^{-1} - 0.40z^{-2} + 0.25z^{-3}} \\
0.25z^{-3}
\end{array}
$$

Hence

$$\text{Output } (z) = 0.37z^{-1} + 0.63z^{-2} + 0.40z^{-3} - 0.25z^{-5}$$
$$+ \ldots$$

The inverse transform of this is thus an output which varies with time, giving the following signal values:

$$f(0) = 0, \ f(1) = 0.37, \ f(2) = 0.63, \ f(3) = 0.40, \ f(4) = 0,$$
$$f(5) = -0.25, \ldots$$

We can determine the output that the system would have when transients have died away by using the final-value theorem.

$$\lim_{t \to \infty} f(t) = \lim_{z \to 1} (z - 1)F(z)$$

Hence, using equation [29]

$$(z - 1)F(z) = \frac{(z - 1)(0.37z + 0.26)}{z^2 - z + 0.63}$$

When $z \to 1$ then $(z - 1)F(z) \to 0$. The steady-state value of the signal is thus 0. Figure 12.20 shows the form of the output signal.

Computer-controlled sampled-data system

Equation [8], earlier in this chapter, is an expression for PID control with a digital computer, namely

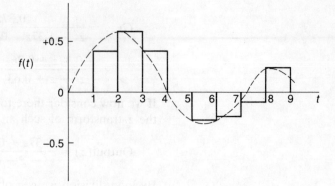

Fig. 12.20 Response of the system in Fig. 12.19 to an impulse input

$$\text{Output} = K_\text{d}e(kT) + K_\text{i}\sum_{0}^{K}\tfrac{1}{2}[e(kT-1) - e(kT)]T$$

$$+ (K_{\text{d}/T})[e(kT) - e(kT-1)]$$

This equation leads to the following z-transform for PID control,

$$\text{Output}(z) = K_\text{p}E(z) + \frac{K_\text{i}T}{2}\left(\frac{z+1}{z-1}\right)E(z)$$

$$+ K_\text{d}\left(\frac{z-1}{z}\right)E(z) \qquad [30]$$

If we have a system, as in Fig. 12.21, which has PI control, then the z-transform of the transfer function of the law being followed by the computer is

$$G_\text{c}(z) = K_\text{p} + \frac{K_\text{i}T}{2}\left(\frac{z+1}{z-1}\right) = \frac{K_\text{p}(z-1) + K_\text{i}T(z+1)}{2(z-1)}$$

With the zero-order hold system, transfer function

$$G_\text{zoh}(s) = \frac{1 - \text{e}^{-Ts}}{s}$$

and the plant with a transfer function of, say,

$$G_\text{p}(s) = \frac{1}{s+a}$$

then

$$G_\text{zoh}(s)G_\text{p}(s) = \frac{1 - \text{e}^{-Ta}}{s(s+a)} = (1 - \text{e}^{-Ts})\left(\frac{1}{as} - \frac{1}{a(s+a)}\right)$$

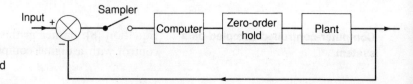

Fig. 12.21 Digital computer-controlled system

The z-transform of the forward path transfer function is thus

$$G_c(z)G_{zoh}(z)G_p(z) = \frac{K_p(z-1) + K_iT(z+1)}{2(z-1)} \frac{(z-1)}{z}$$

$$\left[\frac{(1/a)z}{z-1} - \frac{(1/a)z}{z-e^{-aT}}\right]$$

The z-transform of the closed-loop transfer function is then obtained by substituting the above in

$$G(z) = \frac{G_c(z)G_{zoh}(z)G_p(z)}{1 + G_c(z)G_{zoh}(z)G_p(z)}$$

Example 6

The digital control system described by Fig. 12.21 has a plant with a transfer function of

$$G(s) = \frac{1}{s + 0.1}$$

and is used with a PI digital computer controller with $K_p = 0.9$ and $K_i = 1.1$. What is the z-transform of the transfer function of the system if the sampling period is 1 s?

Answer

$$G_{zoh}(s)G_p(s) = \frac{1 - e^{-Ts}}{s(s + 0.1)} = (1 - e^{-Ts})\left(\frac{10}{s} - \frac{10}{s + 0.1}\right)$$

$$G_c(z) = K_p + \frac{K_iT}{2}\left(\frac{z+1}{z-1}\right) = \frac{1.8(z-1) + 1.1T(z+1)}{2(z-1)}$$

The z-transform of the forward-path transfer function is thus

$$G_c(z)G_{zoh}(z)G_p(z) = \frac{1.8(z-1) + 1.1T(z+1)}{2(z-1)} \frac{(z-1)}{z}$$

$$\left[\frac{10z}{z-1} - \frac{10z}{z-e^{-0.1T}}\right]$$

This can be simplified to give, with $T = 1$ s,

$$\frac{5(2.9z - 0.7)}{(z-1)(z-0.9)}$$

The z-transform of the closed-loop transfer function is then obtained by substituting the above in

$$G(z) = \frac{G_c(z)G_{zoh}(z)G_p(z)}{1 + G_c(z)G_{zoh}(z)G_p(z)}$$

$$= \frac{5(2.9z - 0.7)}{(z-1)(z-0.9) + 0.5(2.9z - 0.7)}$$

$$= \frac{14.5z - 3.5}{z^2 - 12.6z - 2.6}$$

Stability

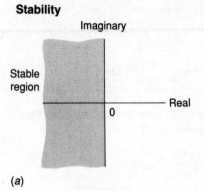

(a)

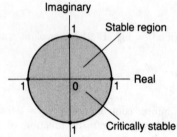

(b)

Fig. 12.22 Stability regions for poles, (*a*) *s*-plane, (*b*) *z*-plane

A linear-feedback system working with continuous signals is stable if all the poles of the closed-loop transfer function lie in the left-hand half of the *s*-plane (see Chapter 8). This means that all the roots of the characteristic equation, i.e. the equation of the denominator of the transfer function expression, must have negative real parts. Since a root *s* can be represented as

$$s = \sigma + j\omega$$

this means that σ must be negative. But equation [12] defines *z* as

$$z = e^{Ts}$$

Hence

$$z = e^{(\sigma + j\omega)T}$$
$$z = e^{\sigma T} e^{j\omega T} \qquad [31]$$

For stability in the *s*-plane we must have σ as a negative number and so σT must be negative. This can only result in $e^{\sigma T}$ having a value between 0 and 1. Since $e^{\sigma T}$ is the magnitude of *z*, i.e. $|z|$, then the *condition for stability* for a sampled system is that all the poles of the *z*-transform of the closed-loop transfer function must lie within a circle of radius 1 (Fig. 12.22). A system is said to be *critically stable* if it has a pole on the circle.

To illustrate this, consider a sampled-data closed-loop system with a transfer function (this was derived earlier as equation [28]),

$$G(z) = \frac{0.37z + 0.26}{z^2 - z + 0.63}$$

This has the characteristic equation

$$z^2 - z + 0.63 = 0$$

The roots of this equation are given by

$$z = \frac{-(-1) \pm \sqrt{(1 - 4 \times 0.63)}}{2}$$

$$= 0.50 \pm j0.62$$

The system is thus stable because the real parts of all the roots are less than 1, i.e. lie within a unit radius circle on the *z*-plane (Fig. 12.23).

The response of a system to an input depends on the locations of the roots. For roots lying within the circle of unit radius an impulse input leads to a response which dies away with time (Fig. 12.24(*a*)(*b*)(*c*)). For roots lying outside the circle of unit radius the response increases with time (Fig.

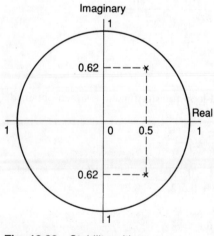

Fig. 12.23 Stability with *z* = 0.50 ± j0.62

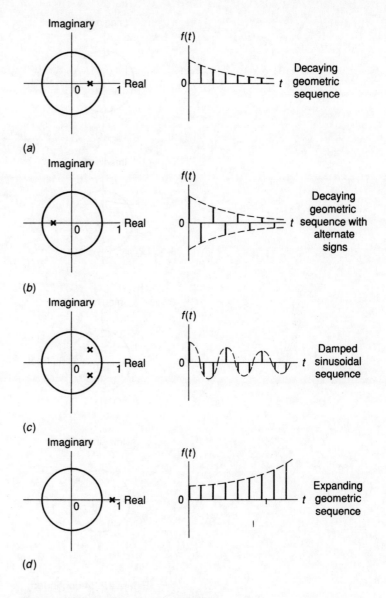

Fig. 12.24 Effect of root location on response to an impulse input

(Figure 12.24 continued on page 310)

12.24(d)(e)). For real roots within this circle (Fig. 12.24(a)(b)) the response is just a steady progressive decay, however for complex roots within the circle (Fig. 12.24(c)) the response is an oscillating decay. For real roots on the circle (Fig. 12.24(f)(g)) the response to the impulse input is a constant output, however for complex roots on the unit radius circle (Fig. 12.24(h)) the response is an oscillation of constant amplitude.

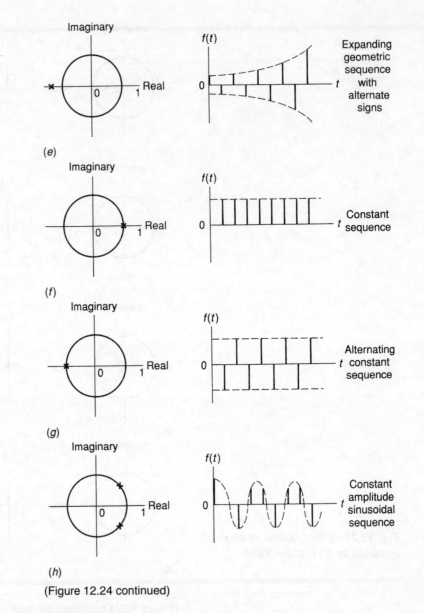

(e)

(f)

(g)

(h)

(Figure 12.24 continued)

Example 7

Determine whether the sampled data systems with the following transfer functions are stable:

(a) $G(z) = \dfrac{5z}{z(z-1)}$

(b) $G(z) = \dfrac{3z-1}{3z^2+2z-1}$

(c) $G(z) = \dfrac{-2z+1}{z^2+2z+1.7}$

(d) $G(z) = \dfrac{-z + 0.6}{z^2 + 0.5z + 0.25}$

Answer

(a) The characteristic equation is

$$z(z - 1) = 0$$

The roots are thus 0 and 1. Since one of the roots lies on the unit radius circle the system is critically stable.

(b) The characteristic equation is

$$3z^2 + 2z - 1 = 0$$

Hence the roots are given by

$$z = \frac{-2 \pm \sqrt{(4 + 12)}}{6}$$

The roots are thus -1.00 and 0.34. The -1.00 root will fall on the unit radius circle and so the system is critically stable.

(c) The characteristic equation is

$$z^2 + 2z + 1.7 = 0$$

Hence the roots are given by

$$z = \frac{-2 \pm \sqrt{(4 - 4 \times 1.7)}}{2}$$

The roots are thus $-1 \pm j0.8$. The roots thus fall outside the unit radius circle and so the system is unstable.

(d) The characteristic equation is

$$z^2 + 0.5z + 0.25 = 0$$

Hence the roots are given by

$$z = \frac{-0.5 \pm \sqrt{0.25 - 1.00)}}{2}$$

The roots are thus $-0.25 \pm j0.43$. The roots thus fall within the unit radius circle and so the system is stable.

Example 8

A sampled-data system is of the form shown in Fig. 12.19, i.e. a zero-order hold system with a plant and unity feedback. If the plant has a transfer function of

$$G_p(s) = \frac{K}{s}$$

for what values of K will the system be stable?

Answer

The zero-order hold system has a transfer function of

$$G_{zoh}(s) = \frac{1 - e^{-Ts}}{s}$$

The forward-path transfer function is $G_p(s)G_{zoh}(s)$ and so

$$G_p(s)G_{zoh} = \frac{K}{s^2}(1 - e^{-Ts})$$

The z-transform of this is

$$\frac{KTz}{(z-1)^2} \frac{(z-1)}{z} = \frac{KT}{z-1}$$

The closed-loop transfer function is

$$G(s) = \frac{G_p(s)G_{zoh}(s)}{1 + G_p(s)G_{zoh}(s)}$$

and thus the z-transform of this is

$$G(z) = \frac{KT/(z-1)}{1 + KT/(z-1)}$$

$$= \frac{KT}{z - 1 + KT}$$

The condition of stability is that the root of z is less than 1. Since the characteristic equation is

$$z - (1 - KT) = 0$$

$$z = 1 - KT$$

If K is restricted to positive values then $|z| = 1$ when $KT = 2$. This is then the critical stability condition. Stability is when $KT < 2$.

The Routh–Hurwitz test for stability

This method of testing whether a system is stable was discussed in Chapter 8. It is a method of determining whether the roots of a system all lie in the left-hand side of the s-plane. This method cannot be directly used with the roots in the z-plane since the condition for stability is that they all have magnitudes within the unit radius circle. However, if z is replaced by

$$z = \frac{1 + W}{1 - W} \qquad [32]$$

then the condition for all the z-roots to lie within the unit radius circle becomes the condition that all the roots of W lie in the left-hand side of the W-plane (Fig. 12.25) and so the test becomes the same format as when it is used with the s-plane.

Consider an example where

$$G(z) = \frac{2z + 1}{z^2 + 0.4z - 0.1}$$

Replacing z by equation [32] gives

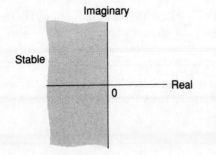

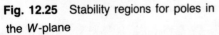

Fig. 12.25 Stability regions for poles in the W-plane

$$G(W) = \frac{2(1 + W)/(1 - W) + 1}{(1 + W)/(1 - W)^2 + 0.4(1 + W/1 - W) - 0.1}$$

$$= \frac{(2 + 2W + 1 - W)/(1 - W)}{[(1 + 2W + W^2) + 0.4(1 - W^2) - 0.1(1 - 2W + W^2)]/(1 - W)^2}$$

$$= \frac{(W + 3)(1 - W)}{1.3 + 2.2W + 0.5W^2}$$

The Routh array can then be constructed for the characteristic equation

W^2	0.5	1.3
W^1	2.2	
W^0	1.3	

There are no changes in sign in the first column and so the poles of W are all in the left-hand side of the W-plane and hence all the poles of z are in the unit radius circle and the system is stable.

Example 9

Is the system giving the following z-transform of transfer function stable?

$$F(z) = \frac{1}{2z^3 + 1.6z^2 + 0.5z + 0.2}$$

Answer

Substituting $(1 + W)/(1 - W)$ for z gives for the characteristic equation

$$2(1 + W)^3 + 1.6(1 + W)^2(1 - W) + 0.5(1 + W)(1 - W)^2 + 0.2(1 - W)^3 = 0$$

$$0.7W^3 + 4.5W^2 + 6.5W + 4.3 = 0$$

The Routh array for this is

W^3	0.7	6.5
W^2	4.5	4.3
W^1	5.8	
W^0	4.3	

Since there are no changes in sign of the entries in the first column the system is stable.

Problems

1 For the continuous time signals shown in Fig. 12.26 state the values of the pulses that would be produced by an ADC up to $k = 3$ if the sampling period T is 1 s.

2 For the continuous time function $3\sin 2t$, sketch the function $f^*(t)$ when the sampling period is 0.2 s.

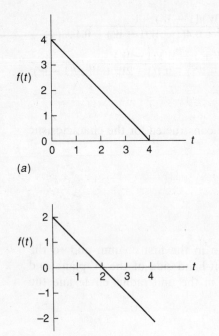

(a)

(b)

Fig. 12.26 Problem 1

3 Explain how an ADC can be regarded as a switch and a DAC as a system with a transfer function of

$$\frac{1 - e^{-T^s}}{s}$$

where T is the sampling period.

4 Explain what is meant by a zero-hold system.

5 Find the z-transforms of the following sequences:

 (a) $f(0) = 1, f(1T) = 0, f(2T) = 0$, and all further values are 0.

 (b) $f(0) = 0, f(1T) = 0, f(2T) = 2$, and all further values are 2.

 (c) $f(0) = 0, f(1T) = 0, f(2T) = 1$, and all further values are 1.

 (d) $f(0) = 0, f(1T) = 1, f(2t) = 2, \ldots f(kT) = k$.

6 Using long division, find the inverse z-transforms in the form of the sample sequence for the following:

 (a) $F(z) = \dfrac{1}{z + 1}$

 (b) $F(z) = \dfrac{2z + 1}{z^2 + 2}$

 (c) $F(z) = \dfrac{2z^2 - 1}{z^2 + 4z + 2}$

7 What are the z-transforms of the following transfer functions?

 (a) $G(s) = \dfrac{1}{s + 1}$

 (b) $G(s) = \dfrac{s}{s + 0.5}$

 (c) $G(s) = \dfrac{1}{s(s + 1)}$

8 A sampled-data open-loop system is of the form shown in Fig. 12.17. What will be the output if the plant has a transfer function of

$$G_p(s) = \frac{s}{s + 0.1}$$

and the input is a unit impulse? The sampling period is 1 s.

9 A sampled-data open-loop system is of the form shown in Fig. 12.17. What will be the output if the plant has a transfer function of

$$G_p(s) = \frac{1}{s^2 + s + 1}$$

and the input is a unit step? The sampling period is 1 s.

10 What is the output of the closed-loop sampled-data system described in Fig. 12.19 when there is a unit step input? The plant

has a transfer function of

$$G_p(s) = \frac{1}{s(s + 1)}$$

and the sampling period is 1 s.

11 A digital computer-controlled system is of the form shown in Fig. 12.21. If the plant has a transfer function of

$$G_p(s) = \frac{1}{s + 0.1}$$

and the computer follows a control law which gives PD control, what will be the transfer function of the closed-loop system $K_p = 10$, $K_d = 4$ and $T = 1$ s?

12 Determine whether the sampled-data systems with the following transfer functions are stable.

(a) $G(z) = \dfrac{4}{z + 0.6}$

(b) $G(z) = \dfrac{2z - 1}{z(z - 1)}$

(c) $G(z) = \dfrac{0.2z - 0.6}{z^2 - 0.1z + 0.5}$

(d) $G(z) = \dfrac{0.1z + 0.7}{z^2 + 0.2z - 0.1}$

13 A closed-loop system consists of a forward path with a zero-order hold system and a plant, transfer function $K/(s + 1)$ and unity feedback. What is the condition for the system to be stable when the sampling period is $1s$?

14 Are the systems giving the following z-transforms of transfer functions stable?

(a) $\dfrac{4(z - 0.1)}{z^3 - 2z^2 + z - 0.4}$

(b) $\dfrac{z^3}{z^3 + 0.3z^2 - 0.2z - 0.1}$

Answers to problems

Chapter 1

1. See text
2. (a) and (b) closed loops with thermostats, (c) open-loop is no thermostat
3. Closed-loop would have feedback to indicate traffic, open-loop would just be time controlled and operate regardless of traffic
4. These might be, though there are other possibilities: Controlled variable − (a) light, (b) temperature, (c) light. Reference value − (a) a set speed/aperture for film speed, (b) set temperature, (c) set minimum light level. Comparison element − (a) differential amplifier, (b) thermostat, (c) a relay. Error signal − (a), (b), (c) the difference between the actual and required values. Control element − (a) the amplifier, (b) thermostat, (c) relay. Correction element − (a) an actuator to change speed/aperture settings, (b) device switching on or varying current through heater, (c) lamp. Process − (a) film in camera, (b) an oven, (c) a particular situation. Measuring device − (a), (c) photocell, (b) thermometer
5. See text
6. (a), (b) two-step, (c) proportional
7. Comparison and control − differential amplifer, correction − relay, process − heater in liquid, feedback via measurement system. Error = difference between the two inputs to the differential amplifier, i.e. reference and measurement voltages
8. $0.14 \, mA/°C$
9. See text: if input multiplied by a constant factor then output multiplied by same factor; if input 1 gives output 1 and input 2 gives output 2 then input 1 plus 2 gives output 1 plus 2
10. (b) 2.0 mA per m/s, (c) 0.6 m/s per mA, (d) 0.3 mA per m/s
11. $-0.2 \, \theta_i$, 4%

12. $-0.90\,\theta_i$, $-6.9 \times 10^{-2}\%$
13. See text
14. Reduced by ½, reduced by $0.5K/(1 + 0.5K)$
15. See text and comparison of open-loop and closed-loop systems

Chapter 2

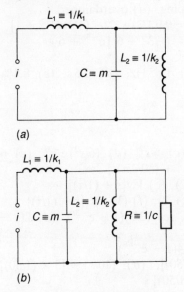

(a)

(b)

Fig. A.1 Chapter 2 problem 6

1. (a) $m(d^2x/dt^2) + c(dx/dt) = F$, (b) $m(d^2x/dt^2) + c(dx/dt) + (k_1 + k_2)x = F$
2. Two torsional spring in series with a moment of inertia block, $T = I(d^2\theta/dt^2) + k_1(\theta_1 - \theta_2) = m(d^2\theta/dt^2) + [k_1k_2/(k_1 + k_2)]\theta_1$
3. $v = v_R + (1/RC)\int v_R dt$
4. $(L/R)\, dv_R/dt + (1/CR)\int v_R dt + v_R = v$
5. $(R_1C)\, dv_C/dt + [(R_1/R_2) + 1]v_C = v$
6. See Fig. A.1
7. See Fig. A.2
8. $RA_2(dh_2/dt) + h_2\rho g = h_1$
9. Same as U-tube example, $p = \rho L\,(d^2h/dt^2) + RA(dh/dt) + 2h\rho g$
10. $RC(dT/dt) + T = T_r$, charged capacitor discharging through a resistor
11. $RC(dT_1/dt) = Rq - 2T_1 + T_2 + T_3$, $RC(dT_2/dt) = T_1 - 2T_2 + T_3$
12. $\Delta F = (2kx_o)\Delta x$
13. $\Delta E = (a + 2bT_o)\Delta T$
14. $\Delta T = (MgL)\Delta\theta$

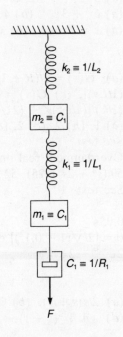

Fig. A.2 Chapter 2 problem 7

Chapter 3

1. See text for examples
2. $dv/dt = 3(V - v)$
3. $4 d\theta_o/dt + \theta_o = 6\theta_i$
4. (a) 57.9°C, (b) 71.9°C
5. (a) $i = (V/R)(1 - e^{-Rt/L})$, (b) L/R, (c) V/R
6. (a) $\tau = 0.6$ s, $G_{SS} = 1$, (b) $\tau = 0.2$ s, $G_{SS} = 0.4$, (c) $\tau = 0.33$ s, $G_{SS} = 0.67$
7. See Fig.3.12, (a) continuous oscillations, (b) underdamped, (c) critical damping, (d) overdamped.
8. (a) 4 Hz, (b) 1.25, (c) $i = \theta_i[(1/3) e^{-8t} - (4/3) e^{-2t} + 1]$
9. (a) 5 Hz, (b) 1.0, (c) $\theta_o = (-32 + 6t) e^{-5t} + 6$
10. (a) 9.5%, (b) 0.020 s
11. (a) 4 Hz, (b) 0.625, (c) 1.45 Hz, (d) 0.5 s, (e) 8.1%, (f) 1.6 s
12. (a) 0.59, (b) 0.87

Chapter 4

1. (a) $6/s$, (b) $(6/s) e^{-3s}$, (c) $6/s^2$, (d) $(6/s^2) e^{-3s}$, (e) 6, (f) $6 e^{-3s}$, (g) $6(50/2\pi)/[s^2 + (50/2\pi)^2]$
2. (a) $1/(s + 2)$, (b) $5/(s + 2)$, (c) $V_0/[s + (1/\tau)]$, (d) $2/s(s + 2)$, (e) $10/s(s + 2)$, (f) $(V_0/\tau)/s[s + (1/\tau)]$
3. (a) $2 e^{-3t}$, (b) $(2/3) e^{-t/3}$, (c) $(2/3)(1 - e^{-3t})$, (d) $2(1 - e^{-t/3})$
4. (a) $(6/5)(1 - e^{-2t/5})$, (b) $4(1 - e^{-t/8})$
5. (a) $[5 \times (1/50)]/s[s + (1/50)]$, (b) $(10/s) + [5 \times (1/50)]/s[s + (1/50)]$, (c) $5/[s + (1/50)]$
6. (a) $5/(s + 2)$, (b) $9/[s(s + 2)] + 5/(s + 2)$
7. (a) 5, 5, (b) 5/2, 0
8. (a) $e^{2t} + 3 e^{-t}$, (b) $4 - 2 e^{-t} - 2 e^{-2t}$, (c) $e^{-t} - e^{-2t}$
9. (a) $x = 2 \cos 8t$, (b) $x = \frac{1}{4} \sin 8t$

Chapter 5

1. (a) $G(s) = 1/(R + 1/Cs)$, (b) $G(s) = 1/[(1/R) + Cs + (1/Ls)]$, (c) $G(s) = 1/(\rho Ls^2 + RAs + 2\rho g)$, (d) $G(s) = (1/k)/[(1/\omega_n^2)s^2 + (2\zeta/\omega_n)s + 1]$
2. (a) 1, (b) 2, (c) 2, (d) 2
3. See text
4. Overdamped, real unequal roots
5. $50/(s^2 + 2s + 25)$, 52.7%
6. See text
7. $t e^{-3t}$
8. 0.05 s
9. $\theta = [2/\sqrt{(1 - 0.1^2)}] \exp(-0.2t) \sin[2\sqrt{(1 - 0.1^2)}t]$
10. $2 e^{-3t} - 2 e^{-4t}$

Chapter 6

1. (a) $2/s(s + 1)$, (b) $20/(s - 1)$, (c) $(s + 2)/(s^2 + 3s + 4)$, (d) $4s/[s^2(s + 1) - 1]$, (e) $2K(s + 1)/(s^2 + 2K - 1)$,

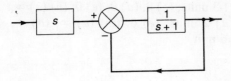

(a)

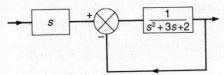

Fig. A.3 Chapter 6 problem 2

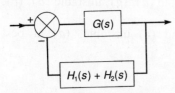

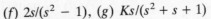

(f) $2s/(s^2 - 1)$, (g) $Ks/(s^2 + s + 1)$

2. See Fig. A.3

3. Transformation 2 gives $\theta_o(s) = G(s)[\theta_i(s) - H(s)\theta_o(s)]$ for a feedback loop. This can be rearranged as $\theta_o(s) = G(s)H(s)[\theta_i(s)/H(s) - \theta_o(s)]$, i.e. a forward path of $G(s)H(s)$ with an input of $[1/H(s)]\theta_i(s)$.

4. See Fig. A.4

5. $\theta_o(s) = \{G_1(s)G_2(s)/[1 + G_1(s)G_2(s)H(s)]\}\theta_i(s) + \{G_2(s)/[1 + G_1(s)G_2(s)H(s)]\}\theta_{d1}(s) + \{1/[1 + G_1(s)G_2(s)H(s)]\}\theta_{d2}(s)$

6. See Fig. A.5. $\theta_r(s) = k_1k_2k_3(1/R_f)G_m(s)/[\tau_1 s + 1 + k_2k_3(1/R_f)G_m(s)]$, where $G_m(s)$ is the transfer function of the motor and is $(1/R_a)k_4(1/c)/[(\tau_2 s + 1)(\tau_3 s + 1) + k_5(1/R_a)k_4(1/c)]$

7. $\theta_o(s)/\theta_i(s) = k_1k_2k_3/s(\tau s + 1)$; with ramp $\theta_o(s) = k_1k_2k_3[1/a^2 s - 1/as^2 + 1/s^3 - 1/a^2(s + a)]$, where $a = 1/\tau$ and so $\theta_o = k_1k_2k_3[1/a^2 - t/a + \frac{1}{2}t^2 - (1/a^2)e^{-at}]$

8. $\omega_o(s) = (1/R_f)k_5(1/c)V/[s^2(\tau s + 1)]$, hence $\omega_o = (1/R_f)k_5(1/c)V[1 - e^{-t/\tau}] = 2(1 - e^{-0.5t})$

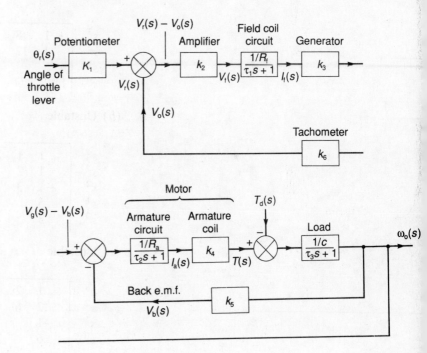

Fig. A.4 Chapter 6 problem 4

Fig. A.5 Chapter 6 problem 6

Chapter 7

1. (a) 0, (b) 0, (c) 1, (d) 1, (e) 2
2. (a) 0, (b) 0, (c) 0, (d) 1
3. (a) 0.82 units, (b) 0.69 units, increasing K reduces the error.
4. (a) 0, (b) ab/K

5. (i) (*a*) 2/7 units, (*b*) 1/3 units, (*c*) 0, (*d*) 0, (*e*) 0; (ii) (*a*) ∞, (*b*) ∞, (*c*) ½ unit, (*d*) 2 units, (*e*) 0; (iii) (*a*) ∞, (*b*) ∞, (*c*) ∞, (*d*) ∞, (*e*) 1/6 unit
6. −1.5 units
7. (*a*) 1, (*b*) 0

Chapter 8

1. (*a*) Stable, (*b*), (*c*) unstable
2. (*a*) Unstable, (*b*), (*c*) stable
3. (*a*) Poles −3, −2, zero ½, (*b*) poles 0, −2, zero ¼, (*c*) poles −2, −3, +4, zeros −2, +3, (*d*) poles (−1 + j1.4), (−1 + j1.4), no zeros, (*e*) poles 0, +3, −5, zeros (−0.5 + j0.87), (−0.5 − j0.87)
4. (*a*) $1/s(s + 1)(s + 2)$, (*b*) $s/(s − 1)(s + 3)$, (*c*) $(s + 1)/(s^2 + 4s + 5)$, (*d*) $(s + 1)(s − 2)/s(s^2 − 6s + 13)$
5. Stable (*d*), (*e*); critically stable (*a*), (*g*); unstable (*b*), (*c*), (*f*)
6. (*d*) can be stable, (*b*) and (*c*) unstable, (*a*) at best critically stable
7. (*a*) Stable

$$
\begin{array}{c|cc}
s^3 & 1 & 8 \\
s^2 & 4 & 12 \\
s^1 & 5 & \\
s^0 & 12 &
\end{array}
$$

(*b*) Unstable

$$
\begin{array}{c|ccc}
s^4 & 1 & 1 & 3 \\
s^3 & 1 & 2 & \\
s^2 & -1 & 3 & \\
s^1 & 5 & & \\
s^0 & 3 & &
\end{array}
$$

(*c*) Unstable

$$
\begin{array}{c|cc}
s^3 & 1 & 2 \\
s^2 & 2 & 6 \\
s^1 & -1 & \\
s^0 & 6 &
\end{array}
$$

(*d*) Stable

$$
\begin{array}{c|ccc}
s^4 & 1 & 24 & 16 \\
s^3 & 4 & 32 & \\
s^2 & 16 & 16 & \\
s^1 & 28 & & \\
s^0 & 16 & &
\end{array}
$$

8. Between 0 and 8

$$
\begin{array}{c|cc}
s^3 & 1 & 4 \\
s^2 & 2 & K \\
s^1 & 4 - \frac{1}{2}K & \\
s^0 & K &
\end{array}
$$

9. One

$$
\begin{array}{c|cc}
r^2 & 1 & -1 \\
r^1 & 2 & \\
r^0 & -1 &
\end{array}
$$

10. K between -2 and 4.5

$$
\begin{array}{c|cc}
s^3 & 1 & 13 \\
s^2 & 5 & 20 + 10K \\
s^1 & 9 - 2K & \\
s^0 & 20 + 10K &
\end{array}
$$

Chapter 9   1. See Fig. A.6, (*c*) $K = 1$, (*d*) $K = 8$

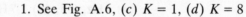

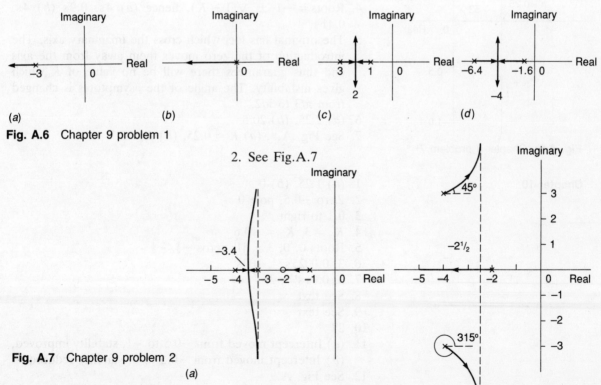

(*a*)   (*b*)   (*c*)   (*d*)

Fig. A.6 Chapter 9 problem 1

2. See Fig. A.7

Fig. A.7 Chapter 9 problem 2

(*a*)

(*b*)

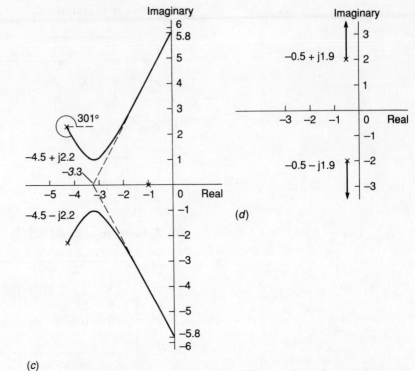

(c)

(d)

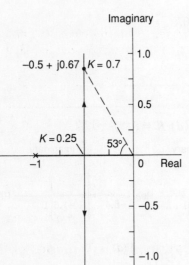

Fig. A.8 Chapter 9 problem 7

3. (a) 4 Hz, 0.5, (b) 3.7 Hz, 0.8, (c) 3.2 Hz, 0.32
4. Roots $= -1 \pm \sqrt{(1-K)}$, hence (a) 4 s, 0.9 s (b) 4 s, 0.41 s
5. The original has loci which cross the imaginary axis. The introduction of the zero moves them away from the axis and thus guarantees there will be no value of K which gives instability. The angle of the asymptotes is changed from $\pi/3$ to $\pi/2$.
6. (a) 2.25, (b) 20.5
7. See Fig. A.8, (a) $K = 0.25$, (b) $K = 0.7$

Chapter 10

1. (a) 1.25, (b) 0
2. Zero -0.5, pole 0
3. 0.1 to right
4. $K_p = 3$, $K_d = -0.6$
5. Roots 0, 0, $3 \pm j1$; zeros $-\frac{1}{2}$, $-\frac{1}{2}$
6. 3, $0.0025\,\text{s}^{-1}$, 100 s
7. 3, $0.01\,\text{s}^{-1}$, 25 s
8. See text
9. See text
10. 27
11. (a) Intercept moved from -0.5 to -1, stability improved, (b) Intercept moved from -0.5 to 0, stability reduced.
12. See Fig. A.9

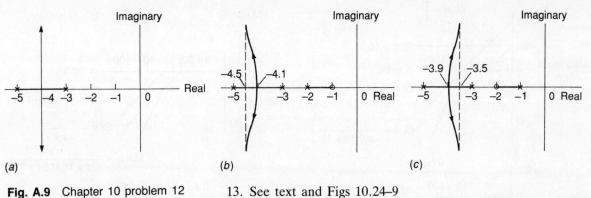

(a)   (b)   (c)

Fig. A.9 Chapter 10 problem 12

13. See text and Figs 10.24–9

Chapter 11

1. (a) $5/\surd(\omega^2 + 4)$, $\omega/2$;   (b) $2/\surd(\omega^4 + \omega^2)$, $1/\omega$;
   (c) $1/\surd(4\omega^6 - 3\omega^4 + 3\omega^2 + 1)$, $\omega(3 - 2\omega^2)/(1 - 3\omega^2)$
2. $\theta_o = 0.56 \sin(5t - 38°)$
3. $\theta_o = 1.18 \sin(2t + 25°)$
4. For magnitude multiply the answers to problem 1(a) and
   (b), i.e. $10/\surd(\omega^2 + 4)(\omega^4 + \omega^2)$. For phase add 1(a) and
   (b) answers, i.e. $(\omega/2) + (1/\omega)$
5.

ω	0	1	2	∞
(a) $\lvert G(j\omega)\rvert$	∞	0.44	0.12	0
ϕ	90°	450°	26.6°	0°
(b) $\lvert G(j\omega)\rvert$	1	0.32	0.16	0
ϕ	0°	−71.6°	−80.5°	−90°

6. See Fig. A.10
7. (a) $100/(s + 1)(0.1s + 1)$, (b) $10/s(0.1s + 1)$, (c) $10s/(s + 1)(s + 0.01)$, (d) $100/(s^2 + s + 1)$
8. See Fig. A.11
9. (a) Always unstable, (b) $K + 1 < (\tau_1 + \tau_2 + \tau_3)(1/\tau_1 + 1/\tau_2 + 1/\tau_3)$, (c) $K > 0$
10. (a) 0.83, (b) 0.66
11. 51°
12. 17 dB, 77°

Chapter 12

1. (a) 3, 2, 1; (b) 2, 1, 0, −1
2. See Fig. A.12
3. See text
4. See text
5. (a) 1, (b) $2z/z^2(z - 1)$, (c) $z/z^2(z - 1)$, (d) $Tz/(z - 1)^2$
6. (a) $f(1T) = 1, f(2T) = -1, f(3T) = +1, f(4T) = -1, \dots$

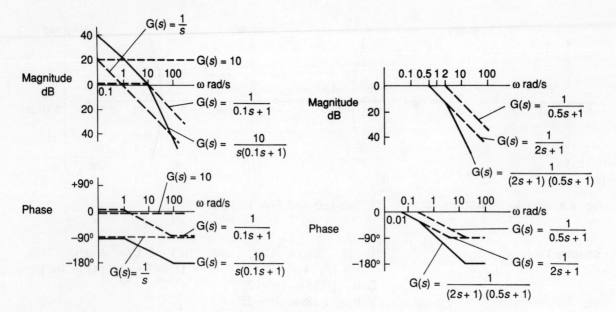

(a)

(b)

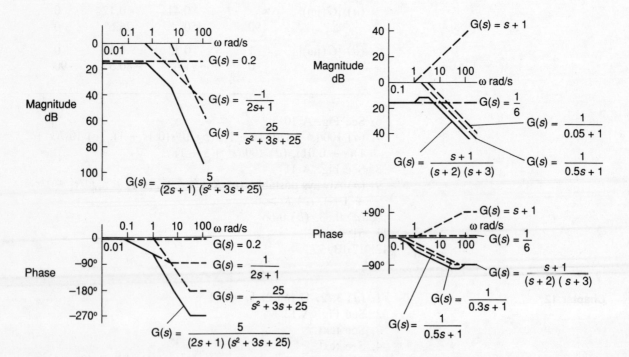

(c)

(d)

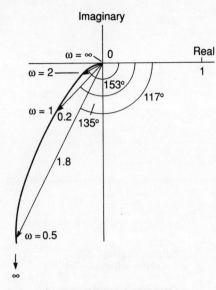

Fig. A.11 Chapter 11 problem 8

(b) $f(1T) = 2$, $f(2T) = -1$, $(f3T) = -4$, $f(4T) = +2$, ... (c) $f(0) = 2$, $f(1T) = -8$, $f(3T) = +27$, $f(4T) = -92$, ...

7. (a) $z/(z - e^{-T})$, (b) $1 - [0.5z/(z - e^{-0.5T})]$, (c) $(1 - e^{-T})z/(z - 1)(z - e^{-T})$

8. $f(0) = 1$, $f(1) = -0.1$, $f(2) = -0.9$, $f(3) = -0.08$, $f(4) = -0.07$, $f(\infty) = 0$

9. $f(0) = 0$, $f(1) = 0.34$, $f(2) = 0.85$, $f(3) = 1.12$, $f(4) = 1.15$, $f(\infty) = 1$

10. $f(0) = 0$, $f(1) = 0.37$, $f(2) = 1.0$, $f(3) = 1.4$, $f(4) = 1.4$, $f(5) = 1.1$, $f(\infty) = 1$

11. $14z - 4/(z^2 + 13.1z - 4)$

12. Stable (a), (c), (d), critically stable (b)

13. $K < 1$

14. (a) Unstable, (b) stable

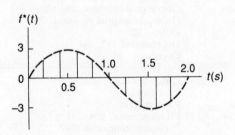

Fig. A.12 Chapter 12 problem 2

Index